IMAGES
of America

GROSSE POINTE FARMS

Lake Shore Road has long been one of the most beautiful coastal drives in America, as this c. 1890 photograph proves. (Courtesy of the Library of Congress.)

IMAGES
of America

GROSSE POINTE FARMS

Terry Nelson and Barbara Nelson

ARCADIA
PUBLISHING

Published by Arcadia Publishing
Charleston, South Carolina

Printed in the United States of America

Library of Congress Control Number: 2023947738

For all general information, please contact Arcadia Publishing:
Telephone 843-853-2070
Fax 843-853-0044
E-mail sales@arcadiapublishing.com

Visit us on the Internet at www.arcadiapublishing.com

*For our Richard, Brownell, and Grosse Pointe South graduate
sons Chris and Jeff and their wives, Katie and Whitney.
For Andrew, Ben, Emmie, Rowan, and Caroline, to whom
we promise to have an endless supply of chocolate balls.
For Linda and Gil, Tim, Toni and Pat, and Tamara,
for always being there. We love you all.*

CONTENTS

Acknowledgments   6

Introduction   7

1.   Sports and Entertainment   9

2.   Grosse Pointe Farms on Lake St. Clair   33

3.   Lakers, Blue Devils, and Knights   55

4.   The Farms Community   93

ACKNOWLEDGMENTS

Very special thanks go to Marica Ostrowski of the Grosse Ponte South Mother's Club for opening the Grosse Pointe South Archives to us, thereby providing dozens of images for our book. Special thanks also go to Gary Hollidge, who seems to know everyone in Grosse Pointe, and to Tim Wittstock, whose personal archives appear to rival those of the school system. Susan Azar and Katie Durno of University Liggett School made available fascinating records and photographs concerning the former Grosse Pointe Country Day School. Another source was Anne Getz Wormer's book *Century: the History of University Liggett School 1878–1978.*

Marcia McBrien, with the assistance of Dianne Romanelli, not only provided great help in telling the St. Paul School story, but also supplied excellent images and information about St. Paul on the Lake Roman Catholic Church. Mary Anne Brush of the Grosse Pointe Public School System provided invaluable images and insights in all things GPPSS, with an assist from Richard Elementary alum James Horan. Likewise, Mark Bolozos of the Grosse Pointe Academy and Derrick Kozicki from the City of Grosse Pointe Farms opened up excellent resources to us as well. Betsy Alexander at the Grosse Pointe War Memorial provided much useful information, as did Madelyn Chrapla from Meadow Brook and also golf historian Anthony C. Gholz Jr.

Great information regarding St. James and St. Paul Lutheran Churches was supplied by Rev. Denise Grant, Richard Allison, and Heidi Korte respectively. Pastor Richard Yeager-Stiver opened up the archives of Grosse Pointe Congregational Church, and Charlie van Becelaere provided a treasure trove of Grosse Pointe United Methodist Church material. Kent Commer found interesting information about the First Church of Christ, Scientist, while Carol Wood and Carol Marks offered outstanding materials regarding Grosse Pointe Memorial Church. Thomas Arbaugh's book *Grosse Pointe War Memorial Church 1865–1990* provided many great images.

In the music department, Tim Shea, Trevor Gates, Tom Weschler, and Bruno Ceriotti of Termoli, Italy, provided great images and insightful information about Grosse Pointe High bands. Thanks also to Mary Lynn Martin and David Lamb of the Grosse Pointe Public Library for coming up with wonderful images and excellent research. Helping to tell the story of the Crescent Sail Yacht Club were Tim Moran and Nick Geisz. Julia Pope, Evan Benn, and Joseph Escribano supplied several outstanding photos of Cottage Hospital. And of course, none of this would have been possible without the support of Jeff Ruetsche and Stacia Bannerman from Arcadia Publishing.

Many of the sources in this volume appear several times, including the Library of Congress (LOC), Grosse Pointe South High School (GPS), Crescent Sail Yacht Club (CSYC), Grosse Pointe Public Library (GPPL), City of Grosse Pointe Farms (GPF), University Liggett (UL) and Conrad Lam Archives, Henry Ford Health System (HFH, LAM Archives).

INTRODUCTION

Indigenous people were fishing and hunting along Lake St. Clair for thousands of years before European visitors arrived. In 1679, René-Robert Cavelier, Sieur de la Salle entered what is now known as Lake St. Clair. That name was given to the waters by Fr. Louis Hennepin, chaplain to the expedition, because their vessel the *Griffin* entered the lake on August 12, the feast day of Sainte Claire. For many years thereafter, French settlers began to inhabit "ribbon farms" along the lakeshore.

In 1775, Capt. Alexander Grant, commander of the British Navy on the Great Lakes, bought a 640-acre farm and built a house known as "Grant's Castle" in what is now Grosse Pointe Farms. Grant's land is said to have been the first cleared land in Grosse Pointe. The name derives from the fine manor house he built on his estate. Grant was lieutenant governor of the county and served on the governor's executive council. He continued to live in Grosse Pointe after the United States took over the Michigan territory in 1796, living there until his death in 1813.

Fr. Gabriel Richard, the namesake of Richard Elementary School in Grosse Pointe Farms, was born in La Ville de Saintes, France, in 1767. After his ordination as a priest in 1790, he emigrated to Baltimore, Maryland, because he refused to declare allegiance to the secular new French Republic. While there, the bishop assigned him to perform missionary work in the northwest territory of the United States, which is now known as the Midwest. His first assignment was as assistant pastor of St. Anne's Church in Detroit. In 1805, after Detroit was devastated by fire, Richard wrote what is now the city's motto: *Speramus meliora; resurget cineribus*, or "We hope for better things; it will arise from the ashes." Richard established a school for Native American and settlers' children, had the first printing press in Detroit, and published periodicals in French and English. Along with Rev. John Monteith and others, he was also instrumental in founding the University of Michigan. Monteith is the namesake of an elementary school in Grosse Pointe Woods.

Sometime after 1850, summer homes were built in Grosse Pointe by D. Bethune Duffield and Francis Palms, part of the beginning of building such homes in the area. In 1879, Grosse Pointe Village was organized and included the property between Fisher Road and Weir Lane, which was just beyond present-day Provencal. In 1889, Grosse Pointe Village extended its boundaries south to Cadieux Road.

A disagreement in 1893 arose within the new village regarding the sale of liquor at a roadhouse called Termont's, which stood on part of the Frederick M. Alger property. This resulted in the organization of two new villages, with what is now the City of Grosse Pointe extending from Fisher to Cadieux and a new town, Grosse Pointe Farms, which took the boundaries of Fisher Road to Weir Lane.

Pictured in this early 1960s photograph are three current or former Grosse Pointe Farms institutions. Looming large at bottom left is Grosse Pointe High School. On the other side of the school parking lot are the former buildings of the Grosse Pointe Country Day School, which at the time of this image were used as an annex to Grosse Pointe High. Just past the annex is Christ Church. (Courtesy of GPS.)

One

SPORTS AND
ENTERTAINMENT

There are plenty of ways to enjoy leisure time in Grosse Pointe Farms. The outstanding Pier Park offers swimming, boating, picnicking, tennis, or simply enjoying the scene from the viewing platform. The War Memorial and its grounds overlooking Lake St. Clair provide a place for reflection. High school and amateur sports are always in the offing, and private clubs offer a variety of activities. And as always, a walk or drive down Lake Shore is hard to beat. Beyond that, over the last century there have been several events in Grosse Pointe Farms, or related to "the Farms," that provided unique opportunities to enjoy time away from everyday life. Shown here is the Grosse Pointe High–based band the One Eyed Jacks performing at one of the many area teen dance venues in 1966. From left to right are Tom Tompkins, Dan Livingstone, and Joe O'Brien. Not pictured is drummer Bill Gross. (Authors' collection.)

Eighteen-year-old Christy Cole Wilson, daughter of Buffalo Bills founder Ralph Wilson and Jane Wilson, had a coming out for the ages when the Supremes performed at her debutante party, held at the Country Club of Detroit (CCD) on June 18, 1965. One can almost hear Diana Ross (left) belting out "Stop! In the Name of Love." Christy can be seen in the crowd just to the right of Ross. (Courtesy of Allyn Baum/*New York Times*/Redux.)

The first Country Club of Detroit building, located in the area of Fisher and Lake Shore Roads, was built in 1886 for an organization called the Grosse Pointe Club. That club was not successful, closing down in 1888, and the building was then used as a summer social recreation center. In 1898, the newly formed CCD purchased the building as its clubhouse. (Courtesy of LOC.)

Pictured on this 1901 National Association of Manufacturers dinner announcement is the newly acquired building of the Country Club of Detroit. The menu that evening included Little Neck Clams; Green Turtle aux Quenellas with Sauterne H&G 1884; Broiled Whitefish, Maitre d'Hotel; Broiled Spring Chicken with G.H. Mumm's Extra Dry; Curacoa Punch; Tomato Farce, Sultana Role, Gauteaux Melange, Fromage, Biscuit and Mocha, and Cigars. (Courtesy of the New York Public Library.)

Over the next few years, the club's board of directors became increasingly concerned over the condition of the current clubhouse. In 1904, the decision was made to build a new one. Noted Detroit architect Albert Kahn was selected to design the new building, pictured here, which opened in 1907. The club's golf links stretched alongside Fisher Road from Lake Shore to Ridge Roads, on land that now contains Richard Elementary and Grosse Pointe South High Schools and the properties in between. (Authors' collection.)

On June 11–13, 1911, an Aviation Meet was held on the golf course of the Country Club of Detroit. The meet was sponsored by the Aero Club of Michigan, led by Fred and Russell Alger, William Metzger, Roy Chapin, Burns Henry, and Alvin McCauley. Attendees were offered the chance to take a spin in Wright Company airplanes. Pictured here is pilot Frank Coffyn (center) with two potential passengers. The newspapers had a field day with Captain Coffyn's surname. (Courtesy of the Smithsonian Institution/National Air and Space Museum Archives.)

The event in itself was newsworthy, but the real news was women taking to the air. One of them was 14-year-old Josephine Alger, making her the youngest person in the country to fly in an airplane. Edward S. George was the first to go up, but soon, many women joined in with the men. Other women taken aloft included Mrs. E.S. Barbour, Mrs. Fred Alger, and Elizabeth Loomis, who exclaimed after her flight, "I wish we could have gone faster!" (Courtesy of the Smithsonian Institution/National Air and Space Museum Archives.)

As the first page of this *Detroit Free Press* sports section shows, the 1915 US Amateur Open, to be held in Grosse Pointe Farms at the Country Club of Detroit, was a very big event. The page featured images of favorites Oswald Kirby, Francis Ouimet, Jerome Travers, and Charles "Chick" Evans. Ouimet, a former working-class caddie, shocked the golfing world in 1913 by winning the US Open at Brookline, Massachusetts. (Courtesy of Anthony Gholz.)

The tournament was played on the newly constructed course designed by legendary golf course architect Harry S. Colt and diagramed in the *Chicago Tribune* just prior to the tournament. At that time, CCD's course stretched from Grosse Pointe Boulevard (now Kercheval Road) to Mack Avenue. Other courses British-born Holt designed included Royal Portrush in Northern Ireland, Royal Lytham and St. Annes in England, and the Eden Course at St. Andrews Links in Scotland. (Courtesy of Anthony Gholz.)

Despite the superstar favorites at the tournament, young Chicago golfer Robert Gardner emerged as the champion. Six years earlier, Gardner had won the amateur tournament at the Chicago Golf Club. At age 19, it made him the youngest person to win the US amateur title. That record stood for 85 years until Tiger Woods won his first amateur championship at age 18. Here, Gardner plays an iron shot to the CCD's 17th green. (Courtesy of *Golf Illustrated* and Yale University.)

In 1954, the US Amateur Championship was again held at the Country Club of Detroit. By this time, the course and clubhouse were at their present locations. A young Arnold Palmer won the championship, defeating runner-up Robert Sweeney at a match that Palmer described as a turning point in his career. After his victory at the country club, he turned pro. Palmer is shown here in 1953, still a member of the US Coast Guard. (Courtesy of the US Coast Guard.)

Detroit Tigers great and Farms resident Ty Cobb was known to have visited the 1915 Amateur Championship to meet with top competitor Francis Ouimet. At the time, Cobb was living nearby on Moross Road. In that year, Cobb set the record for stolen bases with 96, which stood until 1962 when Maury Wills stole 104. That same year, the "Georgia Peach" also won his ninth consecutive batting title, hitting .369. (Courtesy of LOC.)

In more baseball action, in 1961, the year before Wills broke Cobb's stolen base record, the Grosse Pointe Farms Little League Giants won the city championship. Not long after, the Farms Little League program joined with the neighboring City of Grosse Pointe Little League to form the Farms-City Little League organization. Former Major League Baseball player and Farms native Chris Getz played in the Farms-City Little League. Getz is now general manager of the Chicago White Sox. (Courtesy of Tim Wittstock.)

In the late 1920s, several prominent Grosse Pointers decided that there was a need in town for a more exclusive, luxurious movie house. Together, they funded the building of the Punch and Judy Theatre on Kercheval Avenue near Fisher Road, shown here in late 1929 in the final stages of construction. In addition to doormen and plush seats, the new cinema also featured a smoking loge and lounges with fireplaces. (Courtesy of cinematreasures.org.)

Announcing the Gala Opening of the

PUNCH AND JUDY THEATRE

Kercheval and Fisher Road, Grosse Pointe Farms
Wednesday, January 29th, at 8:30

A Real Hollywood and Broadway Premiere

Stage Stars   *Screen Stars*   *Radio Stars*
(All in Person)

Flood Lights   *Red Carpets*   *Celebrities*
Graham McNamee will broadcast direct from Theatre

Seats now on sale at Box Office for Opening Night, $5

As can be seen in this newspaper advertisement, the theater's January 29, 1930, grand opening was meant to rival anything one might see in Hollywood or New York. Part of the hoopla included a nationwide broadcast of the event on NBC featuring announcer Graham McNamee, who is credited with inventing play-by-play sports announcing while calling a World Series game in 1924. (Courtesy of cinematreasures.org.)

The film being shown that evening was *Disraeli*, starring George Arliss in the title role. Also featured were Joan Bennett, Doris Lloyd, David Torrence, and Florence Arliss, wife of George Arliss. The film was nominated for three Academy Awards, with George Arliss winning the Oscar for Best Actor. Admission to the event cost $5 (around $90 today) at a time when movie tickets usually cost a dime. (Courtesy of Warner Bros.)

A Warner Bros. and Vitaphone Picture

In the late 1970s, the operation of the Punch and Judy was taken over by entrepreneurs Lou Bitonti and Larry Lymon. In addition to running wildly successful weekend midnight showings of the *Rocky Horror Picture Show*, Bitonti and Lyman started booking rock music acts, many of them New Wave. A sampling of the acts included the Ramones, Patti Smith, Talking Heads, Joe Jackson, Pat Benatar, and Devo, pictured here at the Punch in 1978. The "P and J" closed in 1984 and is now office space. (Courtesy of Tom Weschler.)

GROSSE POINTE WAR MEMORIAL
YOUTH COUNCIL
MONDAY MOVIE - ROCK

July 22 - "DR. STRANGELOVE"
            CHANCES ARE

July 29 - "THE IPCRESS FILE"
            THE FLOW (FORMERLY THE MIGRAINES)

August 5 - "ZORBA THE GREEK"
            PRESENT TENSE

August 12 - "THE PRIZE"
            SOUNDS OF NIGHT

EVERY MONDAY 7:30 MOVIE
              9:30 DANCE
$1.25 ADVANCE $1.50 AT DOOR

The British Invasion of music in early 1964 led by the Beatles, Rolling Stones, and many other bands led to a tidal wave of American teens forming their own "garage bands." The Detroit area was no different, and Grosse Pointe High became a mixing pot of students looking to find like-minded musicians to form their own bands. Dozens of groups popped up at local dances, as illustrated in this 1968 flier, but three broke through to achieve success. (Courtesy of David Whipple.)

One of those breakthrough Grosse Pointe High–based bands was the Underdogs. Under the guidance of David Leone, owner of the popular dance club the Hideout in Harper Woods, the Underdogs released successful records on Leone's Hideout Records label before being the first white band to sign with Motown. Pictured from left to right are Dave Whitehouse, Tony Roumell, Chris Lena, Michael Morgan, and Steve Perrin. Jack Louisell and Chuck Shermetaro were previous members. (Courtesy of Tomovox.)

Another successful Grosse Pointe band was The Wanted—three of whom attended Grosse Pointe High. Two were from nearby Austin High in Detroit. Their 1967 garage band-style version of "In the Midnight Hour" made it to No. 1 in Detroit and sold well nationally. From left to right are Dave Fermstrum, Tim Shea, Arnie DeClark, Chip Steiner, and Bill Montgomery. (Courtesy of Tim Shea.)

The third Grosse Pointe High band to break out was the Pleasure Seekers. After seeing the Beatles on the Ed Sullivan Show in 1964, friends Patti Quatro, Nan Ball, and Diane Baker decided to form a band. Younger sisters Suzi Quatro and Marylou Ball were asked to join and the band was complete. Their 1965 Hideout single, "What a Way to Die," was featured on an episode of the NBC series *Parenthood* in 2014. (Courtesy of the Patti Quatro archives and Bruno Ceriotti.)

WKNR
MUSIC GUIDE
WEEK OF FEBRUARY 20, 1967
KEENER 13 HITS

1.	HAPPY TOGETHER—TURTLES	WHITE WHALE	(4)
2.	RUBY TUESDAY—ROLLING STONES	LONDON	(2)
3.	YOU CAN TELL ME GOODBYE—CASINOS	FRATERNITY	(7)
4.	KIND OF A DRAG—BUCKINGHAMS	U.S.A.	(1)
5.	STRAWBERRY FIELDS/PENNY LANE—BEATLES	CAPITOL	(28)
6.	KIND OF HUSH/NO MILK—HERMAN'S HERMITS	MGM	(15)
7.	BABY I NEED YOUR LOVIN—JOHNNY RIVERS	IMPERIAL	(5)
8.	LOVE IS HERE—SUPREMES	MOTOWN	(6)
9.	PERSECUTION SMITH—BOB SEGER	CAMEO	(17)
10.	LOVES GONE BAD—UNDERDOGS	V.I.P.	(3)
11.	WHO DO YOU LOVE—WOOLIES	DUNHILL	(8)
12.	MY CUP RUNNETH OVER— ED AMES	RCA VICTOR	(13)
13.	NIKI HOEKY—P. J. PROBY	LIBERTY	(19)
14.	Had Too Much To Dream—Electric Prunes	Reprise	(9)
15.	Hunter Gets Captured—Marvelettes	Tamla	(10)
16.	Go Where You Wanna Go—5th Dimension	Soul City	(11)
17.	Ups and Downs—Paul Revere	Columbia	(20)
18.	Epistle To Dippy—Donovan	Epic	(14)
19.	Indescribably Blue—Elvis Presley	RCA Victor	(22)
20.	For What It's Worth—Buffalo Springfield	Atco	(24)
21.	Hung Up In Your Eyes—Brian Hyland	Philips	(25)
22.	The People In Me—Music Machine	Original Sound	(29)
23.	Dedicated To One I Love—Mamas & Papas	Dunhill	(31)
24.	California Nights—Lesley Gore	Mercury	(30)
25.	I've Been Lonely Too Long—Young Rascals	Atlantic	(26)
26.	Love You So Much—New Colony Six	Sentar	(27)
27.	Jimmy Mack—Martha & Vandellas	Gordy	(KS)
28.	Return Of The Red Baron—Royal Guardsmen	Laurie	(—)
29.	In The Midnight Hour—The Wanted	Detroit Sound	(—)
30.	59th Street Bridge Song—Harpers Bizarre	Warner Bros.	(—)
31.	One More Mountain To Climb—Ronnie Dove	Diamond	(—)

Two Grosse Pointe High–based bands shared the WKNR "Keener Radio" top 31 on February 20, 1967. The Underdog's "Love's Gone Bad" had spent nine weeks on the Keener charts, peaking at No. 2. The Wanted's "In the Midnight Hour" spent five weeks on the WKNR charts, peaking at No. 1. (Authors' collection.)

Both the Underdogs and Pleasure Seekers benefited from associations with local music promoter David Leone. Those bands, as well as other acts including Bob Seger, Ted Nugent, and Glen Frye, played at Leone's very popular Hideout club in Harper Woods, and many recorded songs on his Hideout label. Later on, Leone operated Diversified Management Agency, booking groups around the United States including Aerosmith, Ike and Tina Turner, Chuck Berry, Mitch Ryder, and Iggy Pop. (Courtesy of Marie Anderman and Richard Anderman.)

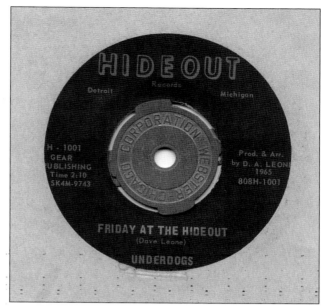

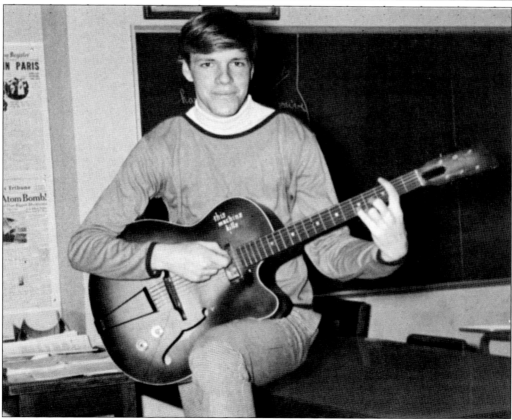

Even Grosse Pointe High foreign exchange student Lee Johnson, from South Africa, got into the music scene by joining a local band. Here, he strums his guitar in 1967. (Authors' collection.)

Suzi Quatro moved to London, England, in 1971 and went on to record three European million-selling records: "Can the Can," "48 Crash," and "Devil Gate Drive." Later, she was tapped by Garry Marshall to play Leather Tuscadero on *Happy Days*. She also appeared in episodes of popular British TV shows *Absolutely Fabulous* and *Midsomer Murders*. This photograph was taken in Detroit in 1978. (Courtesy of Tom Weschler.)

Carrying on the tradition of Grosse Pointe rock stars is Farms native Meg White, one half of the husband-and-wife band the White Stripes. Although her 1996 marriage to the former Jack Gillis lasted only four years, the duo earned six Grammys and sold more than six million albums over their 14-year career. This photograph is from the O2 Wireless Festival at Hyde Park, London, in 2007. (Courtesy of Fabio Venni.)

Grosse Pointe High junior Elsie Scherer is seen here piloting her speedboat in 1955. When she was not racing, Scherer was a home room officer and participated in band, orchestra, the Girls Athletic Association, student assembly finance committee, View Pointe, junior/senior prom committee, and the National Honor Society. (Courtesy of GPS.)

Champion hydroplane racer Lee Schoenith presents trophies to the top women finishers in the 1955 Detroit Marathon powerboat race. To the right of Schoenith, Elsie Scherer proudly holds her trophy, along with Helen Bucurestean (left) and Charlotte Gallagher. Schoenith's family established Detroit riverfront restaurant The Roostertail in 1958, and the business is still family-owned. (Courtesy of boatsport.org.)

Another Grosse Pointe Farms woman to be awarded a boat racing trophy, this time presented by Pres. Calvin Coolidge, was Delphine Dodge, daughter of Horace and Anna Thompson Dodge. The president, in his yacht the *Mayflower*, watched from the Potomac River as Dodge won two straight heats in her boat *Miss Syndicate* to capture the President's Cup in 1927. (Courtesy of schoolfieldcountryhouse.com.)

The September 1933 issue of boating enthusiast magazine *The Rudder* featured this montage of top powerboat racers in the run-up to the upcoming Harmsworth Trophy race to be held on the St. Clair River off of Algonac. Legendary driver Gar Wood, at center, is surrounded by (clockwise from top left) Delphine Dodge, Horace Dodge Jr., George Reis, and Bill Horn. (Authors' collection.)

The previous September, 500,000 spectators lined the shores of Lake St. Clair, including on the Farm's Pier Park, to watch Gar Wood take on Irish-born powerboat racer Kaye Don and his *Miss England III* on a course that followed the Grosse Pointe shoreline. *Miss America X* was the victor in the first heat, and in the second *Miss England III* broke down, handing the race to Wood for his seventh of a total of eight Harmsworth Trophy wins as a driver. (Courtesy of the Reuther Collection, Wayne State University.)

Garfield Arthur Wood, named after two US presidents, poses with his mechanic, Orlin Johnson, in the cockpit of *Miss America X*. In addition to his legendary powerboat racing career, Wood was also an entrepreneur, inventor, and engineer. In 1912, he invented the hydraulic hoist used on dump trucks and at one time held the greatest number of US patents credited to one person. (Courtesy of Hydroplane History.)

Gar Wood's *Miss America X* graced the cover of the September 1933 issue of *The Rudder*. His boat was powered by four 1,600-horsepower Packard engines delivering an unimaginable 6,400 total horses, which according to a Packard ad was three times more powerful than the locomotives that pulled the New York Central Railroad's famed *20th Century Limited* at that time. (Authors' collection.)

In 1932, a local group of former Sea Scouts (part of the Boy Scouts of America) hoped to found a sailing club. An older sailing enthusiast, Chalmer Burns, who was experienced at setting up clubs, helped the young men form their club. With Burns's guidance, the new Crescent Sail Yacht Club (CSYC) was chartered in March 1933. Here, the CSYC holds a regatta in 1942. (Courtesy of CSYC.)

Sometime in the 1940s, this unusual craft headed out from the club into the lake. In December 1941, after the attack on Pearl Harbor, the US Coast Guard Auxiliary requested that a Coast Guard flotilla be formed at CSYC. Joe Vance was made commander, and weekly instructional meetings were held throughout the winter. (Courtesy of CSYC.)

Club members held regular 12-hour watches over Lake St. Clair throughout the war. Here, CSYC members work on their boats next to the boathouse in this undated image. (Courtesy of CSYC.)

Originally, the club was on the Detroit River near Belle Isle. In 1934, the owners of the property announced that the land had been sold and that the club had to leave. Fortunately, Henry B. Joy, president of Packard Motors, saved the day by offering mooring at his peninsula at Kerby and Lake Shore Roads for a nominal rent. The club has resided there ever since. This image is from the 1950s. (Courtesy of CSYC.)

Eventually, after her husband's death, Helen Hall Joy deeded the entire property to the club. There is plenty of racing action in this photograph of a club regatta in 1962. (Courtesy of CSYC.)

Judges observe the 1962 regatta from the observation post atop the boathouse. The boathouse (along with the Joy Bells) was one of the last two remaining elements of the Henry B. Joy estate before it was replaced with a new building in 2005. (Courtesy of CSYC.)

The *Wrangle*, the club's beloved workboat, had its hull delivered in 1940. The engine from the old workboat was rebuilt and installed in the bilge, the deck was laid, and a cabin was made of plywood. After being given a coat of paint, the *Wrangle* was launched. (Courtesy of CSYC.)

The *Wrangle* served the club into the 1980s before finally being retired and dismantled. The boat's transom was saved, and its nameplate now occupies a place of honor on the walls of the clubhouse. (Courtesy of CSYC.)

'56 DODGE Station Wagons

HANDSOMEST WAGONS ON THE ROAD TO SUCCESS

The CSYC was a great setting for several automotive advertisements. In this promotional image, a 1956 Dodge Sierra station wagon uses the CSYC boathouse as the perfect backdrop for the beautiful car. In 1958, an Oldsmobile station wagon was also posed at the club for promotional shots. (Courtesy of CSYC.)

This aerial view of the Crescent Sail Yacht Club was taken around 1940, just seven years after the club's founding and six years after moving onto the Henry B. Joy site. (Courtesy of CSYC.)

This aerial photograph shows the property and harbor of the Crescent Sail Yacht Club today. The original Henry Joy boathouse has been replaced by a beautiful, modern clubhouse. (Courtesy of CSYC.)

The Joy Bells have entertained Farms residents since 1929. They were originally built for Packard Motors president Henry B. Joy for his Grosse Pointe Farms estate Fair Acres, located near what is now the Crescent Sail Yacht Club. After two moves, the Joy Bells are now located at the corner of Grosse Pointe Boulevard and Moross Road. In this image, three Grosse Pointe High students stop to listen in 1963. (Courtesy of GPS.)

Two

GROSSE POINTE FARMS
ON LAKE ST. CLAIR

LIEUT.* A.O. FLUITT (L - R)
PATL. WM. RICE
SERGT.* E. BOYLAN
PATL E.A. BOLO
PAT. W.C. ALLARD
PATL. F.C. CHAMPINE
PATL. ED ALLARD - WAGON
PATL. L. RENO
CHIEF** ELMER BALL
PATL. FRED CHAMPINE
PATL. A. J. ALLARD
SERGT.* J. TROMBLEY
PATL. J. GILCHRIST
LIEUT* J. LEASURE

THIS WAS TAKEN IN REAR OF
HIGHLAND PARK WATER WORKS G.P.F.
MAY 1924

Two publications—one a book, the other a booklet—defined early Grosse Pointe Farms. Michigan historian Silas Farmer's book *Grosse Pointe on Lake Sainte Claire*, written in 1886, was one of the first books to define Grosse Pointe as an entity unto itself. Written as a travelogue, Farmer takes his readers on a virtual journey from Detroit's city center north out to the growing development in the area now defined as the Grosse Pointes. And as it turned out, most of the focus of those places in his book is on present-day Grosse Pointe Farms. The other publication, a booklet entitled *The Village of Grosse Pointe Farms Michigan: Its Past, Present and Future?*, was published in 1922 by the village government as a thought-provoking treatise on the future of the town. In addition to images from these publications, stories of the accomplishments of various Farms residents are also highlighted. This police department photograph from 1924 hangs on the walls of the town hall. (Courtesy of GPF.)

The Village of
GROSSE POINTE FARMS
Michigan
ITS
PAST   PRESENT

AND

FUTURE?

1893-1922

In 1922, the Village of Grosse Pointe Farms marked its 29th anniversary with a 64-page booklet, which honored the past, celebrated the present, and made plans for the future, looking as far ahead as 1950 with growth projections. The front cover is refreshingly not in the style of the usual mundane municipal document. It was clearly meant to inspire curiosity and interest. (Courtesy of GPF.)

Lake Shore Drive on the Curve at the McGraw and Dwyer Estates as it Appeared in 1876.

Looking back 46 years to 1876, the booklet provided an illustration of one of the most iconic views in the Pointes and beyond—looking up Lake Shore toward the curve leading to the drive along Lake St. Clair. (Courtesy of GPF.)

Around 40 years later, two well-dressed pedestrians and three people taking a midday drive enjoy the same classic scene. (Courtesy of LOC.)

The Village Hall.

The Grosse Pointe Farms village hall, opened in 1913, had not changed much by 1922 compared to the same building today. Additions were made in 1919 and 1924, and more renovations and enlargements were done in the 1950s and 1980s. At left, part of the original Kerby School can be seen. (Courtesy of GPF.)

POLICE DEPARTMENT.

Lieut. A.O. Fluitt.   Chief E.C. Ball

Reading from Left to Right, Standing, Bert Easton, Edward Allard, Eugene Bolo, Lawrence Reno, Ross Pursifull, Joseph Leasure, William Rice, Joseph Trombley, Walter Allard; Seated, Sergt. Eugene Boylan, Lieut. Albert O. Fluitt, Chief Elmer C. Ball, Sergt. Louis Wisser.

Members of the Grosse Pointe Farms Police Department proudly posed for their photograph in the booklet. The beginnings of the department were in 1893, when Archibald Michie was appointed the first village marshal. In addition to his police duties, the marshal was charged with collecting the village taxes—a percentage of which paid his salary. (Courtesy of GPF.)

Seen here are the department's Ace motorcycles and an REO patrol vehicle, equipped with a first-aid kit and stretcher. Other equipment included two bicycles, two traffic semaphores with umbrellas, twenty-two traffic signs, ten 12-gauge Winchester shotguns, a rowboat, and a lung motor (resuscitator) donated by Hugo Scherer. (Courtesy of GPF.)

Members of the 1922 Grosse Pointe Farms Fire Department take their turn posing in front of the firehouse doors. According to the brochure, the fire department had two trucks: a Seagrave pumper capable of pushing 750 gallons of water per minute, and a White combination chemical and hose truck. Together, the two trucks carried 3,000 feet of hose. (Courtesy of GPF.)

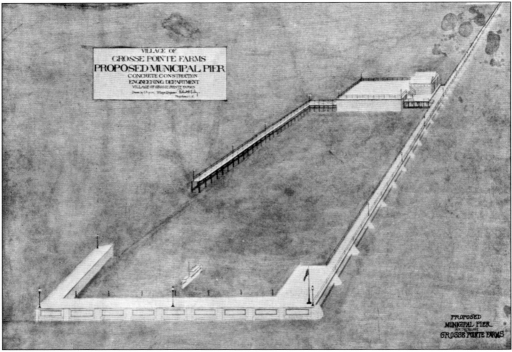

IN CASE OF

FIRE

CALL

HICKORY 625

VILLAGE OF GROSSE POINTE FARMS

Don't get excited. State name and street number clearly.

Keep doors and windows closed to prevent draft.

Don't store oils, paints or grease under stairways.

Don't put hot ashes in wooden boxes or barrels.

For those with telephones, the village provided the fire department's phone number, along with some sound advice. Grosse Pointe Farms was incorporated as a city in 1949. (Courtesy of GPF.)

"There are no public accommodations for bathing or for canoes or other boats. For this reason, plans and estimates were prepared for a municipal pier," stated the village booklet. The park proposal included the building of a 1,000-foot pier at the foot of Moross Road, which could accommodate 36 small boats. Included in the plans were a boat house, a 200-by-300-foot harbor, and a roped-off area for children's bathing. (Courtesy of GPF.)

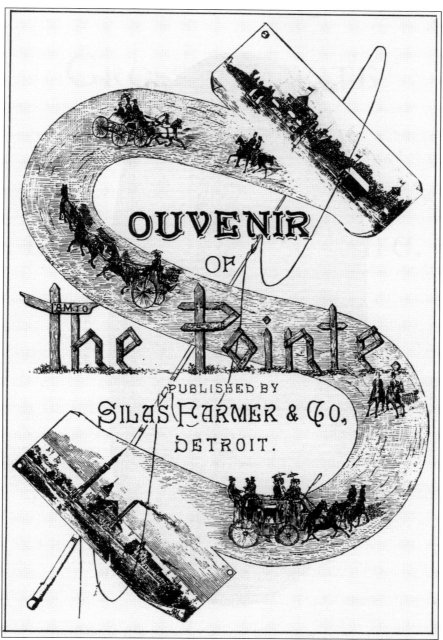

An alternate title for Silas Farmer's 1886 *Grosse Pointe on Lake Sainte Claire* was *Souvenir of the Pointe* (the use of "Souvenir" is intended to mean a guidebook). The title page consisted of this fanciful illustration of a journey which, after beginning at the Michigan Central Station (then on Jefferson Avenue in downtown Detroit), heads north. Shown at bottom is the tower of the waterworks on Jefferson near Belle Isle. The journey then continues out to the shoreline of Grosse Pointe. The scene at the journey's end, from the perspective of Lake St. Clair, is virtually all in today's Grosse Pointe Farms—stretching from the Grosse Pointe Club at left to the tower of W.B. Moran's Bellevue, just past Moross Road, at right. (Courtesy of GPPL.)

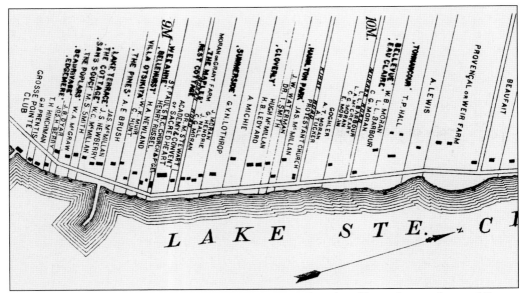

The map contains the following labels (vertical text), left to right:

GROSSE POINTE CLUB · G H PRENTIS · T H HINCHMAN · JOS H BERRY · J B DYAR · W A M°GRAW · "EDGEMERE" · "BEAURIVAGE" · THE POPLARS" M S SMITH · "SANS SOUCI" W C M°MILLAN · J S NEWBERRY · "THE COTTAGE" J S M°MILLAN · "LAKE TERRACE" JAS M°MILLAN · C HUNT · "THE PINES" A E BRUSH · VILLA OTSINITA W K MUIR · H A NEWLAND · HENRY RUSSEL · ST. PAUL'S R.C. CHURCH & P.GR. · ACADEMY & CONVENT OF SACRED HEART · "BELLEHURST" · "WEYANKE" · REST COTTAGE" · "THE MAPLES" J. NAUDRIE · J. MORAN · MORAN GRANT FARM · "SUMMERSIDE" G.V.N. LOTHROP · A MICHIE · "CLOVERLY" HUGH M°MILLAN · H B LEDYARD · DR. I. SMITH · J. W. WATERMAN · JAS. M°MILLAN · "HAMILTON PARK" · PROTESTANT CHURCH · HENRY TUCKER · A T MORAN · GOCHLER · C C MORAN · T. MORAN · C. C. MORAN · L. A. KEARBOUR · MILDRED'S R.C. MORAN · "BELLEVUE" W. B. MORAN · "EAU CLAIRE" L. L. BARBOUR · "TONNANCOUR" T. P. HALL · A. LEWIS · A. LEWIS · PROVENÇAL or WEIR FARM · BEAUFAIT

9M   10M

L A K E   S T E.   C[

This map indicates key landmarks along the way as a visual guide to Farmer's virtual tour of Grosse Pointe. The following images are a sampling of the many homes and landmarks described in his book. Farmer's book was republished in 1974 by the Grosse Pointe Historical Society, preserving this important piece of local history for generations to come. The large "9M" and "10M" at top indicate the distance in miles from Detroit's city hall. (Courtesy of GPPL.)

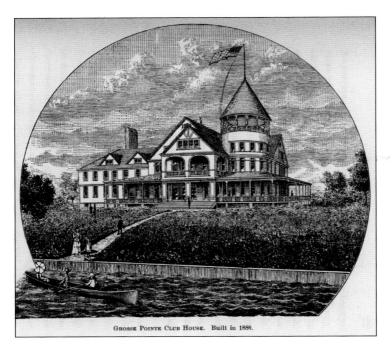

GROSSE POINTE CLUB HOUSE. Built in 1886.

Farmer provides this description of the approach to the first landmark, the Grosse Pointe Club: "And now, through a long avenue of old poplars we see the blue waters of the lake, and, as its red roofs and turrets peep above the trees, we get our first glimpse of the Grosse Pointe Club House." The club was planned to have as many as 300 members; however, after only two years in operation, it closed its doors due to the unreliability of the roads from Detroit. (Courtesy of GPPL.)

RESIDENCES OF JOHN S. NEWBERRY AND JAMES McMILLAN, Grosse Pointe, on Lake St. Clair. Built in 1875.

A little farther along is Lake Terrace, built in 1875 with two homes, one occupied by J.S. Newberry and the other by James McMillan. Farmer writes, "The owners of Lake Terrace were the first to build costly houses at the Pointe, and their success in constructing elegant, graceful and convenient country seats, induced others to follow their example . . . with their neighbor, Mr. Brush, they built the long dock which stretches out into the lake." This image is from an earlier Farmer book. (Courtesy of LOC.)

Right at the bend in the road on the map on the previous page is St. Paul's Roman Catholic Church and parsonage. Farmer writes, "The religious interests of residents are cared for by two churches. "The worshipers at the old French Catholic Church of St. Paul are especially numerous. By the way, why should it be called St. Paul rather than St. Peter, who was the patron of fishermen?" (Courtesy of GPPL.)

ST. PAUL'S ROMAN CATHOLIC CHURCH.

ACADEMY OF THE SACRED HEART.

Next door is the Academy & Convent of Sacred Heart. Farmer explains, "This institution is one for the education of young ladies. The Academy is one of the most complete of the kind in the country, the building cost nearly one hundred thousand dollars ($3.2 million today), and is furnished with every modern convenience. It is four stories in height, and is heated by steam throughout. . . . Pure lake water is supplied by a steam engine." (Courtesy of GPPL.)

"SUMMERSIDE." Residence of G. V. N. Lothrop. Built in 1850.

Of the second lot past the academy, Farmer announces, "We have now reached 'Summerside,' the residence of George V. N. Lothrop, at present United States Minister at St. Petersburg, Russia. Rare specimens of old trees, and orchards of the finest fruits adorn the grounds. Mr. Lothrop was impressed with the beauty of the Pointe soon after his arrival in Detroit, and purchased 130 acres there in 1850." Lothrop originally was from Connecticut. (Courtesy of GPPL.)

Taking out a small portion of James McMillan's property near the lake is the Protestant Church. As Farmer describes it, "There is also a pretty little Protestant church conducted by a few of the most enterprising residents. It is a church of all evangelical creeds and is attended on Sundays by most of those whose time on that day is not devoted to the worship of nature." That little church was the foundation of today's Memorial Church. (Courtesy of GPPL.)

THE PROTESTANT CHURCH. Built in 1876.

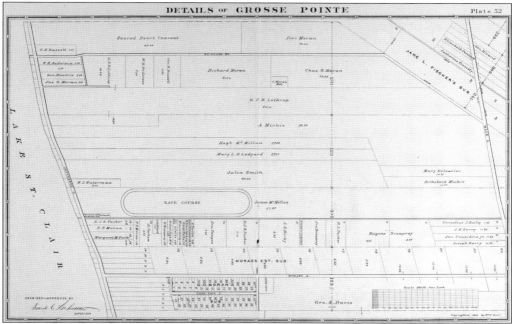

Hamilton Park is identified as a "race course" on this 1893 map. Farmer does not provide an image of this property, but does provide this description: "The adjoining grounds, known as Hamilton Park, are the property of James McMillan, who has converted the fifty acres into a pleasant driving park. When the main road is in bad order, the fast 'nags' of the Pointe are exercised in Hamilton Park." Present-day Voltaire Place and Vendome Road were once part of the track. (Courtesy of LOC.)

"TONNANCOURT." Residence of T. P. Hall. Built in 1880.

Two lots past Morass (now Moross) Road is Tonnancour, the home of Thomas P. Hall. Farmer describes some of the entertainments to be had there: "On the lake side is a spacious boat-house built in the Swiss style of architecture. Here dancing parties, theatricals and other evening amusements take place." Of course, Hall's boathouse was more than a place for recreation. Like many other Grosse Pointe summer residents, Hall's yacht would be used for commuting to Detroit as much as pleasure cruising. (Courtesy of GPPL.)

Not far beyond Tonnancour is the Provencal-Weir house. Built around 1823, not only is it the oldest house in the Farms, it is also the oldest building in all the Grosse Pointes. Silas Farmer commented, "The Provencal house standing on front of the farm, near the lake, is a good example of the old-time French home." It was moved to its present location on Kercheval Avenue in 1914 and is now the home of the Grosse Pointe Historical Society. (Authors' photograph.)

The Restful Car

"The supreme combination of all that is fine in motor cars."

Prestige ʻʻ The Packard owner, however high his station, mentions his car with a certain satisfaction—knowing that his choice proclaims discriminating taste as well as a sound judgment of fine things.

For the Packard is one of the world's few fine cars universally approved by the enthusiastic owners of other famous makes.

Recognized everywhere, as supremely typifying America's genius for perfection

in things mechanical, Packard cars go further in possessing to a marked degree that subtle attribute—prestige.

Packard prestige, sensed if not defined by every Packard owner, is reflected in the car's aristocratic beauty, its distinction, its luxury and comfort, its superb performance—unexcelled in traffic or on the open road.

PACKARD

A S K T H E M A N W H O O W N S O N E

For years, Packard was among the world's most prestigious automobile manufacturers. The company's confident slogan, "Ask the Man Who Owns One," announced that this was the car to drive. The first Packard automobiles were produced in 1899 when entrepreneurs James and William Packard teamed up with George Lewis Weiss in Warren, Ohio. Four hundred Packard cars were built in Warren between 1899 and 1903. When Grosse Pointe Farms resident Henry B. Joy bought one in 1902, he was so impressed with the vehicle that he convinced fellow Farms residents Truman H. Newberry and Russell A. Alger Jr. to refinance the company and move it to Detroit. James Packard was chosen to serve as president until his retirement in 1909. (Wikimedia Commons.)

Packard investor Russell A. Alger Jr. was born on February 27, 1870. His father, Russell Alger, was a Civil War hero, governor of Michigan, US senator, and member of Pres. William McKinley's cabinet. Like his father, the younger Russell was in the lumber business. In 1896, Alger was married to Marion Jarves. In 1910, the family moved into their home on Lake St. Clair, which they called the Moorings. (Courtesy of LOC.)

Packard's slogan may have been "Ask the Man Who Owns One," but Mrs. Maurice "Honey" Nelson of Detroit is also justly proud as she poses by her new Packard Deluxe in 1942. The World War II–era windshield sticker with the letter "A" indicates that the car's owner was entitled to four gallons of gasoline per week. (Authors' collection.)

Packard investor Truman Newberry and his wife, Harriett, are seen here attending an unknown event around 1910. Truman was born on November 5, 1864. After graduating from Yale in 1885, he became the manager of the Detroit, Bay City & Alpena Railway until 1887. Following that, he was the president of Detroit Steel and Spring Company. He served in the Spanish-American War and was a US senator from 1919 until 1922. (Courtesy of LOC.)

Packard power is on full display as champion speedboat racer Gar Wood, on the left, and his mechanic Orlin Johnson prepare to fire up the four Packard 1600-horsepower V-12 engines on *Miss America* X in the early 1930s. (Authors' collection.)

Henry Joy became Packard's president in 1909. Under his leadership, Packard began to expand its reputation for technology and luxury. Joy was a flamboyant executive who in moments of impatience would frequently say, "Let's do something, even if it's wrong!" He also was the first president of the Lincoln Highway Association. Joy recruited municipal and state participation and even determined much of the route itself. He served as chairman of Packard until 1926. (Courtesy of LOC.)

In 1941, the US Army Air Corps changed engines in its P-51 Mustang fighter, seen here over Ramitelli, Italy, in 1945, from Allisons to faster Rolls-Royce Merlin engines. Packard was given the license to build them in the United States and eventually built over 54,000. Even though Packard's financial situation was strong after the war, it could not compete with the Big Three of General Motors, Ford, and Chrysler. The last Packard rolled off the line in 1958. (Courtesy of LOC/Toni Frissell.)

John and Horace Dodge, seen in the back seat, were born in Niles, Michigan. From an early age, they exhibited excellent mechanical abilities. They began their careers supplying parts to Ransom E. Olds and later to the Ford Motor Company. In 1914, they began producing their own automobiles, focusing on innovations such as a speedometer, electric ignition, and an all-steel body. (Courtesy of LOC.)

Pictured is a Dodge magazine ad from the 1920s. Horace and Anna Thompson Dodge settled in Grosse Pointe Farms at Lake Terrace on Lake St. Clair. John and Matilda Rausch Dodge planned to build a mansion on Lake Shore in the Farms, but in 1920, the post–World War I flu epidemic claimed both brothers, and the home was never completed. Matilda Dodge married Albert Wilson and settled at Meadow Brook in Rochester. In 1940, Matilda Wilson was appointed lieutenant governor of Michigan, the first woman in the United States to hold that office. (Authors' collection.)

Another automotive giant living in Grosse Pointe Farms was Henry Ford II, seen here at Schiphol Airport in the Netherlands in 1963. Between 1945 and 1980, Ford II held the offices of president, CEO, and chairman of the board of directors. Under his watch, he took Ford public, oversaw the building of the Glass House—Ford's new world headquarters—and led the development of the Mustang, creating a whole new class of cars. (Courtesy of Hugo van Gelderen.)

In the early 1960s, Henry Ford II entered into negotiations with Enzo Ferrari to buy his company, hoping to get a foothold in racing—including the 24 hours of Le Mans race. After those negotiations failed, Ford II decided to build his own high-performance endurance car, resulting in the GT40, which won at Le Mans in 1966, 1968, and 1969. Pictured is a GT40 at Germany's Nurburgring road course in 1965. (Courtesy of Lothar Spurzem.)

Grosse Pointe Farms resident Roy D. Chapin, seen here, was the leader of a group of local business leaders who founded the Hudson Motor Car Company in 1908. The car was named after Detroit retailer Joseph L. Hudson, who provided the majority of the company's capital. A Chapin-led subsidiary, Essex Motors, developed the first mass-produced affordable enclosed car. The popularity of this vehicle caused the industry to move away from open touring cars toward all-weather autos. (Courtesy of LOC.)

Swedish tennis stars Torsten Johansson (left) and Lennart Bergilen pose with a 1947 Hudson at a company-sponsored tennis open in Stockholm. In looking for markets overseas, Hudson hoped to find an answer to the growing dominance of the Big Three. Unfortunately, like Packard, it was not to be, and the last Hudson was built in 1957. (Wikimedia Commons.)

In 1889, New York–born Hamilton Carhartt, pictured here, began a business in Detroit making tough, high-quality overalls for railroad workers. He grew up in Jackson, Michigan, where he married Annette Willing in 1881. By 1910, the company had grown to include several plants in the United States, Canada, and Great Britain. With the Great Depression taking a toll on its core business, Carhartt expanded its lines to include work clothes for farm and ranch workers. The Carhartts had three children: Hamilton Jr., Wylie, and Margaret. The younger Hamilton and brother Wylie both moved to Grosse Pointe, with Wylie becoming company president in 1937. Lifelong Farms resident Gretchen Carhartt, daughter of Wylie, in recent years became the public face of the company. Not only did she serve as chairman emeritus, she was also active in the local jazz scene, including founding a record company, opening a jazz bar, and supporting the Detroit Jazz Festival. Below is a c. 1904 company logo. (Left, courtesy of Bassmaster; below, courtesy of the *Tacoma Times*.)

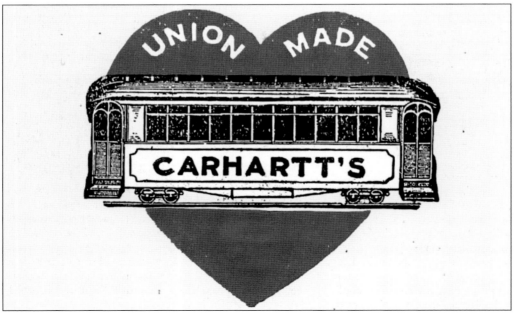

At one time, Detroit had three major newspapers: the *Detroit News*, the *Detroit Free Press*, and the *Detroit Times*. The *Times* was founded in 1900 by James Schermerhorn as the *Evening Times*. The paper eventually went into receivership, and in 1921 was purchased by newspaper magnate William Randolph Hearst. In 1934, Grosse Pointe Farms resident William E. Anderman became the publisher of the *Times*, a post he held until 1957. Anderman's beginning in newspapers was as a carrier for the *Chicago Daily News*. He came to Detroit in 1921 to serve as Hearst's advertising director, general manager, and then publisher. During his tenure, circulation reached over 436,000 in 1951. Eventually, the *Times* could not keep up with its competitors, and in 1960, it was absorbed by the *Detroit News*. (Courtesy of Richard Anderman/LOC.)

Living next to the Andermans on Lake Shore was Max McKee, president of Sands Products Corporation. Sands had an interest in shipping companies, including the Wisconsin and Michigan Steamship Company. McKee wanted to build an "ocean liner for the Great Lakes," and when a retired World War II–era merchant ship, the *Marine Star*, became available, he found his ship. The ship was purchased by McKee's company in 1951, and the work of conversion to a passenger ship began in 1953. (Courtesy of Linda and Gil Missana.)

After two years and $8 million, the conversion was complete. In 1956, *Aquarama*, at 520 feet the largest passenger ship ever to ply the Great Lakes, began touring. In 1957, regular service between Detroit and Cleveland began. The ship carried both passengers and cars, and the one-way trip took six hours. The beauty and luxury of the ship are evident in this photograph. Although very popular, the ship was unprofitable, and service ended in 1962. (Courtesy of John Rochon and the Sarnia Historical Society.)

Three

LAKERS, BLUE DEVILS, AND KNIGHTS

In 1868, the Academy of the Sacred Heart opened a free school for the children of St. Paul Parish, followed in 1885 by the Sisters of the Sacred Heart opening their own building nearby on Lake St. Clair. In 1914, the Grosse Pointe Country Day School was founded in the city of Grosse Pointe but soon moved to a campus on Grosse Pointe Boulevard in the Farms. The Grosse Pointe Public School System was formed in 1922 when schools from five separate rural districts (Trombly, Cadieux, Kerby, Vernier, and Cook) were combined into one unified district. In 1927, the Parish of St. Paul opened its own school, and in 1969, the Grosse Pointe Academy replaced the Sacred Heart Academy. Seen in the mid-1960s, this Quonset hut was initially given to the Grosse Pointe Country Day School in 1945 by steel manufacturer George Fink. After Country Day moved out in 1954, its buildings went to the Grosse Pointe school district and the Quonset hut was used as a driver's education classroom. It was dismantled in the early 1970s. (Courtesy of GPS.)

The beginnings of the Grosse Pointe Country Day School were in 1914, when due to a whooping cough epidemic, a number of Grosse Pointe parents whose children attended either Liggett or the Detroit University School wished for their children to be educated closer to home. Eventually, a new school called the Grosse Pointe School was inaugurated. While a campus on Grosse Pointe Boulevard was being prepared, the students attended classes at a home on Roosevelt Place in Grosse Pointe, pictured here. (Courtesy of UL.)

In February 1916, the new campus (seen here in 1930) opened, housing 98 pupils and employing 15 teachers. The goal of the school at that time was to prepare students to attend eastern secondary schools such as Phillips Exeter in New Hampshire or the Groton School in Massachusetts. The house on Roosevelt Place was used to provide housing for teachers during a World War I housing shortage and was sold in 1918. (Courtesy of UL.)

The Nursery School

The Nursery School occupies the clubhouse on our Cook Road property. These ten acres—one acre of which is composed of woods—afford a perfect opportunity for the young child to work and play away from the older children in an atmosphere of beauty and harmony. Here the child learns his first lessons in cooperation and in conforming to certain simple standards of good behavior.

Listening attentively to stories helps to lengthen the child's span of attention.

By 1941, Grosse Pointe Country Day School had evolved to become its own college prep school. In that year, the school published an illustrated booklet highlighting the excellent programs offered. The school was organized into nursery, lower, intermediate, and upper schools. Here, the nursery school is highlighted. (Courtesy of UL.)

SCHOOL ★

The Lower School develops rhythmic expression through bodily response to music.

Even the youngest share the time of our expert physical education instructors.

The third grade Social Studies projects take them to many lands.

The lower school consisted of kindergarten through third grade and was in a separate unit apart from the intermediate and upper schools. (Courtesy of UL.)

SCHOOL ★

Inter-School competition is based on equality.

A pause for instruction — vital in all play.

Stories become real through dramatization.

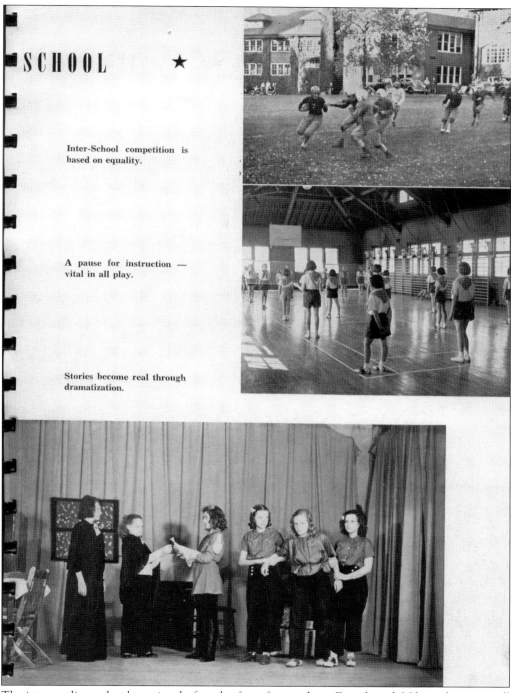

The intermediate school consisted of grades four, five, and six. Fourth and fifth grades were still largely self-contained, while sixth grade began to prepare students for the full departmentalization of the upper school. (Courtesy of UL.)

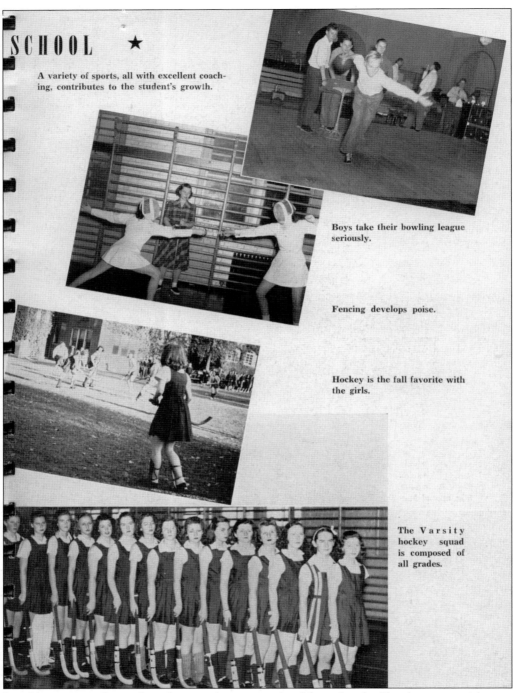

SCHOOL ★

A variety of sports, all with excellent coaching, contributes to the student's growth.

Boys take their bowling league seriously.

Fencing develops poise.

Hockey is the fall favorite with the girls.

The Varsity hockey squad is composed of all grades.

In addition to traditional academics, the upper school (seen here) offered many enrichment opportunities in the curriculum including debate, radio arts, and photography ("from tintypes to colored movies"). Dramatics including operettas, musical comedies, pageants, and seasonal plays written and directed by the students were another part of the curriculum. (Courtesy of UL.)

Exemplifying the dramatics curriculum is this all-female enactment of *Hamlet*, performed in 1930. The cast, in alphabetical order, was made up of Eugenie Carhartt, Constance Harry, Jane Peabody, Marie Sanger, Beth Standart, Louise Stockard, Betty Tant, Phoebe Walker, Mary Warwell, and another unidentified student. (Courtesy of UL.)

Tenth graders take a few moments to relax and socialize between classes in this 1942 photograph. Clockwise from lower left are Fred Farr, Jeanne Bradley, Barbara Burrows, Margaret McKean, Bill Duffield, Peckie McMahon, Lydia Kerr, Carlisle Frost, Barbara Hughes, Mary Joyce Malow, and Julie Harris. Standing in back are, from left to right, Bob Den Uyl, Jim May, and Helen Livingstone. (Courtesy of UL.)

Field hockey stars Sally Whitehead, Julie Harris, Barbara Hughes, and Barbara Bayne sport their letter sweaters with the Grosse Pointe School initials in this early 1940s photograph. Throughout its time in the Farms, this was commonly used in place of the school's formal name. (Courtesy of UL.)

Julie Harris, class of 1944, went on to much bigger things. She was a five-time Tony-winning and three-time Emmy-winning actress who was also twice nominated for a Best Actress Oscar. One of those nominations was for her role as Abra, playing opposite James Dean in *East of Eden*. (Authors' collection.)

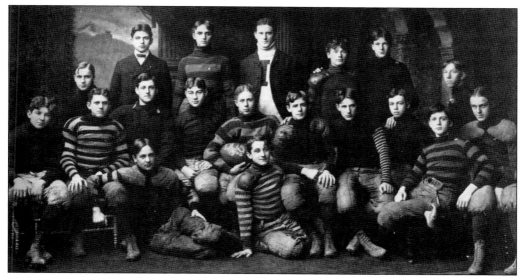

University Liggett School was formed from the heritage of three independent schools. One was the Detroit University School (DUS), founded in 1889 and represented here by the 1900 football team. Only a few are identified, including Coach Herbert Barbour, at center in the back row, and to his immediate right, Leffingwell Whiting, who went on to captain the Brown University baseball team. Sitting to the right of the player holding the football is Trafton Keena, who was chosen "the best DUS all-around school boy." (Courtesy of UL.)

G. Mennen Williams, pictured with Israeli prime minister David Ben Gurion, was a Grosse Pointe Farms native who spent his early years at DUS. He went on to attend Salisbury School in Connecticut before going to Princeton. Williams was elected Michigan governor for six consecutive two-year terms between 1948 and 1960 and later was elected to the Michigan Supreme Court. DUS and Grosse Pointe Country Day merged in 1954 to form the Grosse Pointe University School (GPUS). (Courtesy of UL.)

The third, and oldest, school that makes up today's UL was originally called Detroit Home and Day School, founded by Rev. James Liggett in 1878. The school's name was changed to the Liggett School in 1912. Seen here is the first graduating class of Detroit Home and Day—the class of 1881. They are Olive Farrand, Jessie Van Court, Nellie McMillan, Margaret Inglis, Grace Whitney, and Florence Ives. (Courtesy of UL.)

In 1969, Liggett merged with GPUS to form University Liggett School. The home of the Knights, UL today occupies a campus on Cook Road in Grosse Pointe Woods. It is a preschool/kindergarten–12th grade college-prep school. Its founding mission is "to help young people more completely realize that which is within them and in doing so make positive contributions to society." (Courtesy of UL.)

The establishment of the Grosse Pointe Academy in 1969 makes it the newest school—created out of the oldest school—in Grosse Pointe Farms. In 1969, the Sisters of the Sacred Heart deeded their school and buildings to a new board of trustees. The original 1885 Sacred Heart Academy's charter was then transferred to the new school. The new board incorporated the school as an independent nonprofit coeducational elementary school. The school's main classroom building, seen here, dates from the late 1920s. (Courtesy of the Grosse Pointe Academy.)

The Grosse Pointe Academy chapel dates from 1899. It contains classic period details such as wood paneling, wood carvings, and beautiful stained-glass windows. (Courtesy of the Grosse Pointe Academy.)

Mr. John M. Poplawski —
Headmaster

School Directors — Miss Viviano,
Mrs. Deutsch & Mr. L. Reeside

Starting a new school from scratch is a tall order. This task was successfully undertaken by this group of dedicated school leaders, as well as many others, pictured here in 1971. (Courtesy of the Grosse Pointe Academy.)

The Grosse Pointe Academy eighth-grade class of 1989 looks happy, eager, and ready to move on to their secondary school experience, wherever it may be. (Courtesy of the Grosse Pointe Academy.)

How proud the original Sisters of the Sacred Heart would be to know that well over a century after its construction, this building is still serving the youth of Grosse Pointe Farms and beyond. The 1920s classroom building is visible to the left. In 1987, the building was listed in the National Register of Historic Places. (Authors' photograph.)

In 1868, the Academy of the Sacred Heart opened a free school for the children of St. Paul Parish in a four-room building. In 1926, the parish decided to build its own school, which opened in 1927. Pastor Alonzo Nacy, seen on the cover of this newsletter, proudly exclaimed that St. Paul was "a parish school second to none in the state or, for that matter, the nation." (Courtesy of St. Paul School.)

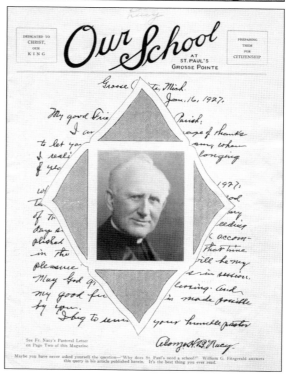

The St. Paul seventh-grade class of 1926 contained many familiar names, including a Beaufait, a Newberry, a Kerby, a Webber, and four Beaupres. Father Stack is on the left, and Father Nacy is on the right. (Courtesy of St. Paul School.)

By 1951, parish growth made additional space necessary. In that year, famed architect Minuro Yamasaki was engaged to build a new classroom building. Yamasaki is most known for designing the original World Trade Center in New York. Seen here is the architect's rendering of the proposed building. (Courtesy of St. Paul School.)

Pictured is Yamasaki's St. Paul classroom building today, as clean and fresh as it was in the 1950s. Other notable Yamasaki buildings include the Wayne State University School of Education building, the Dhahran Air Terminal in Saudi Arabia, Century Plaza Hotel in Los Angeles, and the Michigan Consolidated Gas Company in Detroit. (Authors' photograph.)

Coach Ed Lauer twice led St. Paul to the Class C basketball state championship—in 1961 and 1964. Here, the 1964 champs happily surround the coach after their 53-39 victory over Byron Center. The Lakers' Bob Martin scored 23 points while Jimmy Bigham added 21. (Courtesy of St. Paul School.)

CO-CAPTAINS: Mary Allor, Marilyn Sutherland
FRONT ROW: Sharron Ducastel, Mary Margaret Van Damme, Mary Allor, Coach Kay Dungan, Marilyn Sutherland, Connie Molitor, Donna Mayo, Janet Wrubel. BACK ROW: Lou Ann Moxley, Joan Hock, Mary Ellen Beaupre, Blanche Finney, Charlotte Klein. Janet Hock, Kathryn Andre.

Junior Varsity City Champs

FRONT ROW: Barbara Unti, Marilyn Blondell, Nancy Mason, Coach Kay Dungan, Marilyn D'Hooghe, Barbara Wines, Joan Marks.
BACK ROW: Beverly Cadieux, Joyce Zemper, Lynn Van Tiem, Joan Heldt, Mary Sutherland, Claire Lenz.

The year 1952 was a very good year for St. Paul's girls' basketball teams. With varsity winning the Catholic Youth Organization crown, its third straight, and junior varsity taking the city championship title, "We Made It 'Two' in '52" became the victory cry. (Courtesy of St. Paul School.)

Janet Kingsbury and Gerald Lubeski seem to be very much enjoying the music of Ray Gorrell's band at their senior prom, held at the Lochmoor Country Club in 1951. (Courtesy of St. Paul School.)

Basketball coach Ed Lauer sits at the wheel of this 1960 Chevrolet station wagon supplied by the now-closed Ver Hoven Chevrolet dealership. There is plenty of room in the wagon for St. Paul varsity football players; from left to right are Neil Endres, Frank Cobb, Bob Lauer, Ray Martin, David Barlow, and Robert Colosanti. St Paul closed its high school in 1971 but still offers kindergarten through eighth-grade education. (Courtesy of St. Paul School.)

In 1944, the Grosse Pointe Board of Education adopted a postwar district-wide building schedule. The applications submitted to the Michigan Planning Commission included plans for a new Kerby Elementary School, replacing the original Kerby School built in 1905. The new school was to be located on a vacant site owned by the school district on Kerby Road. In 1949, the new Kerby School, seen here, opened. (Courtesy of Tim Wittstock.)

Kerby teacher Nancy E. Dilloway's 1962–1963 sixth-grade class happily smiles for their group portrait. The principal at that time was Dr. Custer Homeier. Dr Homeier was a teacher at Pierce Junior High School before becoming Kerby's principal, a post he held for many years. (Courtesy of Tim Wittstock.)

Richard Elementary School was the last of four Grosse Pointe public schools to be built in the 1920s, although it actually opened in early 1930. Its French Renaissance Revival architecture was meant as an homage to the area's early French settlers and specifically Fr. Gabriel Richard, a much-loved local Roman Catholic priest who also served one term in the US House of Representatives. (Courtesy of the National Register of Historic Places.)

On a sunny day in 1952, happy young children file out of Richard Elementary after a day of learning. The interior of the school is as impressive as the exterior, featuring marble-lined hallways, Pewabic tile fireplaces, and fish ponds. In 1994, Richard was listed in the National Register of Historic Places. (Courtesy of otisourcat/flickr.)

"We Name a School" was a skit performed by Eudora Snyder's seventh-grade homeroom after Brownell Junior High, which was a school within a school at Grosse Pointe High, was formally named in 1941. Tommy Ronan played Dr. Brownell in the skit, and he is seen here admiring the painting of his character, which hung in the cafeteria. (Courtesy of Tim Wittstock.)

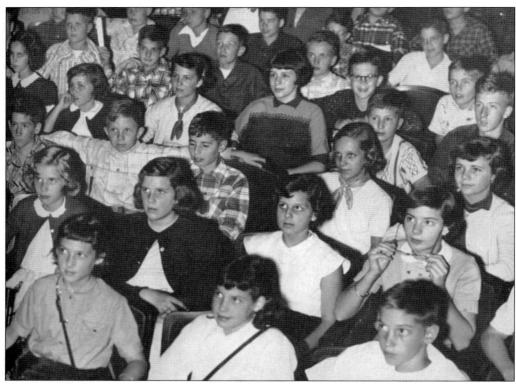

Everyone at this Brownell assembly in the mid-1950s is paying at least some degree of attention, except for the jovial lad wearing glasses staring directly at the camera. (Courtesy of GPS.)

Seventh-graders Martin Hutchinson, Judie Limbrock, Maralyn Winter, and Howard Frieze are seen here in chorus class. If they chose to pursue music further, a glee club and an a cappella choir were open to them. (Courtesy of GPS.)

Seventh-grader Jim Benedict is intently watching the conductor during the first public appearance of the Brownell Band in the early 1940s. (Courtesy of Tim Wittstock.)

Finally, in 1958, Brownell Junior High had a home of its own. The land on which the school is built was previously occupied by Battery A of the US Army's 99th AAA Gun Battalion. Over time, 500 soldiers were assigned to tend the four 90-mm anti-aircraft guns housed at the site. (Courtesy of Tim Wittstock.)

Everyone in the Brownell orchestra is locked in as they prepare for an upcoming performance in 1966. (Courtesy of Tim Wittstock.)

Construction of the new high school for Grosse Pointe began in January 1927 on a site that included the former golf course of the Country Club of Detroit. Construction proceeded rapidly, and by late 1927, the building was nearing completion, with only its 134-foot tower left to finish. (Courtesy of GPPSS.)

JUNE
1942

VIEW POINTE
GROSSE POINTE HIGH SCHOOL

The cover of the June 1942 yearbook, published just six months after the attack on Pearl Harbor, conveys a somber tone. Concern, vigilance, and determination were the watchwords of virtually all endeavors in the United States during those uncertain and dangerous days. Up until the early 1960s, students would begin their schooling in either September or February depending upon their birthdate. Accordingly, there were two graduating classes each year, in January and June, and so two yearbooks were published. (Courtesy of GPS.)

In those unsettled days, when there was no guarantee that bombs would not fall, everyone did their part. Here, physics teacher William A. Mann and his student aide, senior Fred Schriever, familiarize themselves with their assigned lookout position in the school's tower in case of an air raid. (Courtesy of GPS.)

Throughout World War II, no one could be certain that they were safe from attack by air. Therefore, every precaution was taken to be prepared for such an event. At Grosse Pointe High, there was room enough for the entire student body to be seated on the first floor, as seen in this 1942 photograph. In case of fire, all students would be evacuated to the basement of the industrial arts building. (Courtesy of GPS.)

Benjamin R. Marsh Jr., a 1934 Grosse Pointe High graduate, joined the US Navy in August 1940 after attending the University of Michigan. Attaining the rank of ensign, in November 1941 he was assigned to the USS *Arizona*, which was stationed at Pearl Harbor. On December 7, 1941, the *Arizona* was destroyed and sunk by enemy bombs. Over 1,100 crewmen died, including Ensign Marsh. Like most of his shipmates, his body has never been recovered. (Courtesy of GPS.)

On September 25, 1943, Ensign Marsh's mother, Marjorie Bills Marsh, christened the USS *Marsh*, named for her son. Built at the Defoe Shipbuilding Company of Bay City, Michigan, the *Marsh* (DE 699) was a Buckley-class destroyer escort and took part in several convoy missions in the Atlantic and then was assigned to accompany many more convoys in the Pacific. (Authors' collection.)

The new building, designed by local architect George J. Haas, opened on February 1, 1928. Haas also designed Grosse Pointe's Defer and Mason Elementary Schools as well as Hamtramck High School in addition to his many other east side projects. This image from 1958 perfectly captures the serene beauty of the high school as students enjoy their lunch break courtesy of the open campus policy. (Courtesy of GPS.)

The chandeliers add a special charm to what was then the high school library as students pursue their studies below in this early 1950s photograph. At this time, there were over 19,000 books and thousands more photographs, pamphlets, and recordings to enrich student learning. (Courtesy of GPS.)

Senior Mary Dahlen takes a few moments to read on the library balcony in 1954. Behind her is one of the murals depicting the history of education painted in the 1930s by Detroit artist Edgar Yeager as a Works Progress Administration project. After suffering damage over the years, in 1990, Yeager was hired to restore the paintings. (Courtesy of GPS.)

The octagonal Bert H. Wicking Library took the place of the original library in 1964. Although still the library, in 1971, classrooms were built encircling the octagon with a structure simply called the S-building. Immediately beyond the Wicking, facing Grosse Pointe Boulevard, stands what was called the Annex, the former buildings of the Grosse Pointe Country Day School. The Annex had been used for many years as auxiliary classrooms for the high school. (Courtesy of GPS.)

Grosse Pointe High principal Walter R. Cleminson looks on as student association president Fred "Skip" Pessl receives the Crusade for Freedom "freedom scroll" in 1950. Cleminson served as principal from 1940 until his untimely death in 1957. Cleminson Hall was named in his honor after the Wicking Library opened. Pessl went on to graduate from Dartmouth. He became a noted geologist, and later a writer. (Courtesy of GPS.)

Instructed by Mr. and Mrs. Robert Parzych, these summer school students are learning their keyboarding skills, 1960s style. Note the microphone at bottom center, apparently needed by the instructors to address the students over the clatter of the typewriters. (Courtesy of GPS.)

Art students, such as these, have the advantage of attending class on the school lawn on fine days, such as this one in 1964. (Courtesy of GPS.)

Senior business-class student Mercedes Marco has been learning the latest in 1953 business technology. Surrounding her, counterclockwise from the upper left, are a typewriter, dictaphone, comptometer (the first successful key-driven mechanical calculator, patented in 1887), calculator, adding machine, and a bookkeeping machine. (Courtesy of GPS.)

The founding of the Grosse Pointe South Mothers Club coincided with the opening of the school in 1928, and it has been a central part of school life since then. The organization provides activities such as the senior all-night party and scholarships for deserving students. Pictured here are the club's tea chairwomen from 1961. From left to right are Mrs. P. Gould, Mrs. R. Nigro, Mrs. P. Nash, Mrs. H. Biers, Mrs. R. Eddy, Mrs. E. Rushmer, Mrs. F. Adams, Mrs. R. Thumann, Mrs. J. Cornelius, Mrs. E. Evenden, and Mrs. K. Davies. (Courtesy of GPS.)

In another year this architect's drawing will have become a reality in steel and brick. Costing close to a million dollars, it will give Grosse Pointe one of the best high school gyms in the state. Every need and activity of the community has been considered in the layout of the Auditorium-Gymnasium.

By the 1950s, the high school's original gym had become outdated. A new combination gymnasium and auditorium, as seen in this original architect's drawing, was planned to be built by 1955. (Courtesy of GPS.)

The finished product is seen here in 1955, with newly painted walls, a freshly varnished floor, and the game clock ready to time its first basketball game. (Courtesy of GPS.)

The newly designed Border City League (BCL) flags for the first time adorn the gymnasium walls in the late 1950s. After the BCL disbanded, South joined the Eastern Michigan League (EML). In 1988, the EML disbanded and all the teams, including South, joined the Macomb Area Conference. (Courtesy of GPS.)

Newly designed Border Cities League flags, only recently hung in the auditorium-gymnasium, symbolize the closer cooperation among neighbors that is sure to characterize the future.

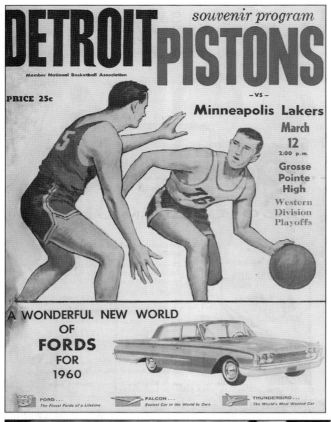

On March 12, 1960, the Detroit Pistons were scheduled to host the Minnesota Lakers in a first-round NBA playoff game. Normally, the Pistons would have played at Olympia Stadium, but it was booked due to a previously scheduled Ice Capades show. The teams had played a preseason exhibition game at Grosse Pointe High, so the decision was made to play there again. Lakers superstar Elgin Baylor scored 40 points, and the Pistons lost a nail-biter 113-112. (Courtesy of Bill Dow.)

In 1962, the Grosse Pointe High boys swim team won the state championship. Here, Blue Devil swimmers Don DeMeulemeester and Barry Brink get off to a good start against two unseen competitors from Monroe in the 200-yard freestyle event. (Courtesy of GPS.)

State Champs 1954

TOP ROW
Christo, Lamb, Ogden, Whittingham, Muir, Leamon, Ireland, Wood,
Russell, Wolfe, Swindlehurst, Nicholson, Bell

3rd row
End Coach Baver,
Worley, Stewart,
Mulliken, Scheverle,
Genova, Jenkins,
De Meulenaere,
S. Moceri
Line Coach Kurvink

2cd row
Mgr. Groschner,
Morse, Seroki,
Horn, Billups,
Barry, Kingsbury
Hooker, Duerksen
Baver, C. Moceri,
Wyse, Walker
Mgr. Thomas

1st row
Head Mgr. Passanante, Turk, Jeffries, Herschelman, Underwood, Eugenio,
Lineberger, Head Coach Wernet, Dow, Prince, MacMillan, Wassall, Anderman,
Bremer, Mgr. Donovan

During the 1954 football season, head coach Ed Wernet had led the Blue Devils to a 7-0 record. Port Huron, favored by many to become state champions, was the final opponent. The Grosse Pointers went on to a decisive 56-12 victory over the Big Reds, culminating in a perfect 8-0 season. As a result, they were crowned the consensus all-state champions for 1954. (Courtesy of the Grosse Pointe South Gridiron Club.)

The Girls Athletic Association (GAA) was formed in the 1890s nationally across high schools and colleges to promote physical activity among girls and young women. Participants could earn points for participation leading to the awarding of the GAA badge, which was the equivalent of the boys' varsity letter. These students were named Outstanding Senior Girl Athletes for the January 1952 graduating class. Displaying their GAA badges are Alice James, Emily Harding, Loa Zay Sheridan, Carol Fredericks, and Nancy Ramsey. (Courtesy of GPS.)

High school girls did not have the opportunity to play for state championships, as there were no equivalent sports opportunities for them until after the federal Title IX act in 1972. There were opportunities for girls' athletics though, through the GAA, intramural, or informal games. These Grosse Pointe High students are ready for a game of football. (Courtesy of GPS.)

Once Title IX became law, it took only four years for Grosse Pointe South girls to chalk up their first state championship. Under the guidance of coach Stephanie Prychitko, the South girls won the 1976 state Class A tennis championship—and went on to win it every year until 1986. Pictured are the 1977 state champs. From left to right are (first row) Jennifer Transue, Lori Wood, Kim Pierce, Lisa Micou, and Pam Pierce; (second row) Coach Prychitko, Lee Robinson, Jennifer Tewes, Melinda Manos, Barb Warren, and Carolyn Reisig. (Courtesy of GPS.)

The 1939 baseball team poses in front of the school doors. From left to right are (first row) Art Wittstock, Lawrence Shock, Bruce Boxstang, Dick Bodycomb, Bob Hood, Bob Carman, Dan Corin, and Bill Allard; (second row) ? Barnes, coach ? Peterson, Blair Martin, ? Cobbs, Bill Schmicht, Joe Faust, Horse Abrambs, Bill Williams, Russ Davis, Ken Needham, ? Bennett, coach Forest Geary, and Warren Piche; (third row) Eddie Smith, David La Vine, Tom Braden, and two unidentified. Under coach Dan Griesbaum, South baseball won state championships in 2001 and 2018. (Courtesy of Tim Wittstock.)

Paul A. Rehmus, the Grosse Pointe High principal from 1937 to 1940, said, "Whether a team loses or wins, the student body in Grosse Pointe stands firmly behind it. Our students attend the games for the sheer enjoyment of seeing athletic prowess displayed in the spirit of fine sportsmanship." (Courtesy of Tim Wittstock.)

This Grosse Pointe High cheerleader is doing her best in exhorting the fans to show their spirit at a home football game in 1951. (Courtesy of GPS.)

Here is an example of the finest in prom wear in 1953. Smiling for their pre-prom picture are, from left to right, seniors Jane Condon and Marty McCormick with juniors Judy Wilson and Jack Adams. (Courtesy of GPS.)

As the top four academically in the class of 1951 and members of the National Honor Society, Tom Lamb, Ellen Vander Brug, Patricia Marx, and Diana Nylund stand here literally and figuratively at the top of their class. (Courtesy of GPS.)

Michigan governor George Romney addresses fellow participants at the Grosse Pointe High football field after a civil rights march on June 29, 1963. The march, sponsored by the NAACP, began in the village and then proceeded down Kercheval Avenue to the Hill, ending up at the football field. Five years later, Dr. Martin Luther King Jr. made his famous speech at Grosse Pointe High School. (Courtesy of Reuther Collection, Wayne State University.)

A historic occasion in the history of Grosse Pointe South High School occurred on March 14, 1968, the night Martin Luther King came to Grosse Pointe. King was invited to speak about "The Other America" by the Grosse Pointe Human Relations Council. Although the planned speech caused much controversy in the community, most people in the crowd of 2,700 listened respectfully. Three weeks later, on April 4, Dr. King was assassinated in Memphis, Tennessee. (Courtesy of Gary Hollidge.)

The June 1956 Grosse Pointe High graduating class was to be the first not to have their commencement on the front lawn, as seen in this image from a previous year. Instead, it was held in the new gymnasium. Fortunately, over time commencement would once again take place on the lawn, as is the tradition today. (Courtesy of GPS.)

Four

THE FARMS COMMUNITY

The Grosse Pointe Farms community is made up of many remarkable institutions that serve the mind, the body, and the soul. The Henry Ford Medical Center–Grosse Pointe, long known as Cottage Hospital, has afforded outstanding medical care to Grosse Pointers for over a hundred years. The Grosse Pointe Public Library system, with its central branch in the Farms, has reliably provided up-to-date knowledge and learning opportunities since 1928. And spiritually, Grosse Pointe Farms churches have fed the souls of area faithful at least since 1835, when St. Paul Parish was organized to serve Grosse Pointe Township. This picture from 1964 shows a Cottage Hospital nurse standing ready to receive her newest patient. (Courtesy of HFH, LAM Archives.)

THE NAMING OF LAKE ST. CLAIR.

The second centennial anniversary celebration of the naming of Lake St. Clair took place at Grosse Pointe, Michigan, on August 12, 1879. The first page of the printed circular programme of the centennial exercises contained a cut of the Griffin and the following announcement:

FRANCE, 1679.   UNITED STATES, 1879.

THE GRIFFIN.

AUGUST 12, 1879.

SECOND CENTENNIAL CELEBRATION OF THE ORIGINAL NAMING OF LAKE ST. CLAIR,

WHICH TOOK PLACE ON THE

Twelfth day of August, 1679, on board of the Schooner "Griffin," of Grosse Pointe, by Robert Cavelier Sieur de la Salle.

REGATTA WITH ACQUATIC SPORTS.

Yachts will start from the pier at Grosse Pointe promptly at 2 P. M., to be followed by other "Ancient Mariners" in their "Bumboats."

A short-lived local newspaper, the *Detroit Post and Tribune*, filed this report in August 1879: "August 12, 1879 was the two-hundredth anniversary of the discovery of Lake St. Clair by Robert Cavelier, le Sieur de la Salle, commander of the Griffin, the first vessel that ascended the Detroit River. The second centennial of any notable event is a rare thing in this new country, and so . . . the residents of Grosse Point (sic) resolved to celebrate the occasion." The festivities took place on the property of Alfred Brush on Lake St. Clair not far from St. Paul church. This was one of the first major celebrations to occur in what is now Grosse Pointe Farms. The day's festivities included a regatta followed by, according to a contemporary circular, music, prayer, speeches, a specially written song, fireworks, and a "chorus by the Grosse Pointers." (Courtesy of Canadiana.)

In 1865, a group of local like-minded individuals met to draw up the Articles of Association of the Protestant Evangelical Association of Grosse Pointe. The aim of the group was to build a "place of worship on the banks of Lake St. Clair" for the Protestants of Grosse Pointe Township. The result was this Gothic-style wooden church on the corner of Kerby and Lake Shore Roads, which opened its doors for worship in 1867. (Courtesy of Grosse Pointe Memorial Church.)

In 1892, church trustees decided to build a new church on recently acquired land—the site of the current church today. Trustees Henry Ledyard, Joseph Berry, and Truman Newberry engaged Detroit architects Mason and Rice, who designed the Grand Hotel on Mackinac Island, to draw up the plans for the new church. The "ivy covered church," as it came to be called, was dedicated on Sunday, July 1, 1894. (Courtesy of Grosse Pointe Memorial Church.)

In 1920, the church was reorganized as the Grosse Pointe Presbyterian Church. In 1921, a new pastor, Rev. Dr. George Brewer, seen here, advocated for a new church to accommodate the changing face and the increased economic growth of Grosse Pointe. The congregation agreed, and soon plans were made for a new church. (Courtesy of Grosse Pointe Memorial Church.)

Truman Newberry, pictured here with his wife, Harriett Barnes Newberry, was named board president and given the job of overseeing the project. In 1924, Newberry indicated that he and his family had plans to give a memorial church to Grosse Pointe. The Grosse Pointe Presbyterian Church's name would be changed to Grosse Pointe Memorial Church in honor of his and his brother John's parents. (Courtesy of Grosse Pointe Memorial Church.)

96

In February 1926, the plans for the new church were finalized. It would be built in modern English Gothic style with a 78-foot tower featuring eight bells to be cast in Croydon, England. The completed church was dedicated on May 15, 1927. Membership increased from 180 in 1920 to 950 in 1930. (Courtesy of Grosse Pointe Memorial Church.)

The beautiful interior of Grosse Pointe Memorial Church is on full display in this photograph of a wedding around 1940. (Courtesy of Grosse Pointe Memorial Church.)

The church provided many opportunities for fun and social interaction, including having its own basketball team. Here is the 1948–1949 team. From left to right are (first row) Jim Farquhar, Harold Lee, Dan Cronin, Bruce Bockstanz, and Marvin Asmus; (second row) Bob Knell, Bruce Reynolds, John Nelson, Arthur Dannecker, and Gorden LaRue. (Courtesy of Grosse Pointe Memorial Church.)

The church also showed movies on Friday nights. In 1927, attendance averaged 360 per show. Charging 15¢ for children and 25¢ for adults, the church cinema had a profit of $643 that year, about $11,000 today. There were other entertainments as well. Neta and Rudy Hirt are all smiles as they prepare to put on a show in 1929 given by the church's young married couples group. (Courtesy of Grosse Pointe Memorial Church.)

St. Paul Parish was established in 1835 to serve the French settlers living on Lake St. Clair. The original log building was replaced in 1848 by the building pictured here. After the present church was built, it was used as a parish hall until 1914, when it was dismantled. (Courtesy of St. Paul on the Lake.)

Rev. John F. Elsen was the prime mover behind changing the face of Grosse Pointe Farms forever. As St. Paul's pastor from 1889 to 1898, he was the driving force behind the beloved Gothic structure facing Lake St. Clair—a landmark from land or sea. This iconic church, which seats over 600, replaced the 1848 church. Unfortunately, Elsen did not live to see the church's 1899 opening, having passed away earlier that year. (Courtesy of St. Paul on the Lake.)

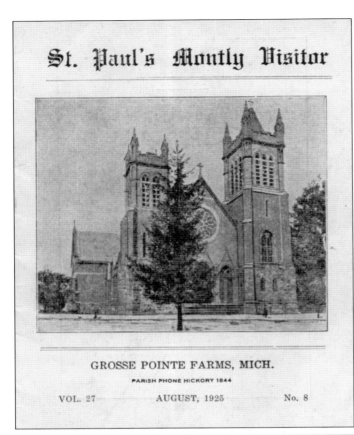

St. Paul's Montly Visitor

GROSSE POINTE FARMS, MICH.

PARISH PHONE HICKORY 1844

VOL. 27   AUGUST, 1925   No. 8

Father Elsen's legacy, St. Paul church, graces the cover of this church newsletter. The newsletter was instituted by Father Elsen's successor, Fr. Alonzo Nacy, soon after his appointment. (Courtesy of St. Paul on the Lake.)

Father Stack (left) and Father Nacy pose with altar boys in the mid-1920s. In addition to leading his flock over his 30-year pastorate, Nacy oversaw the building of a school, convent, and rectory. (Courtesy of St. Paul on the Lake.)

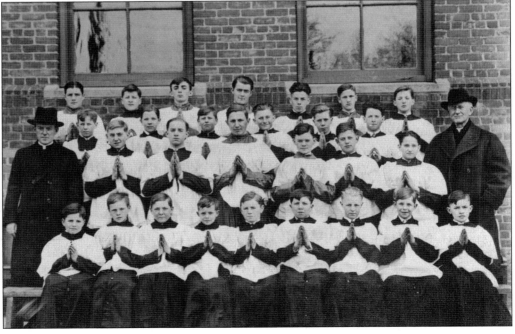

St. Paul's bright, inspirational interior was the perfect setting for this early 1960s wedding. (Courtesy of St. Paul on the Lake.)

The architect of St. Paul, Harry J. Rill, was born in Germany in 1854 and emigrated to the United States in 1881. Around 1894, Father Elsen chose him to design the beautiful church building that overlooks Lake St. Clair today. Pictured here, St. Paul Boy's Choir members pose on the church steps in 1959. (Courtesy of St. Paul on the Lake.)

On June 9, 1978, St. Paul parishioners, as well as all of Grosse Pointe, held their collective breath as the church was hit by a devastating fire. The blaze, reportedly caused by defective wiring, caused serious damage to the roof, altar, and other interior areas. Fortunately, the quick response of the local fire departments avoided worse damage, but restoration still took over a year to complete. (Courtesy of St. Paul on the Lake.)

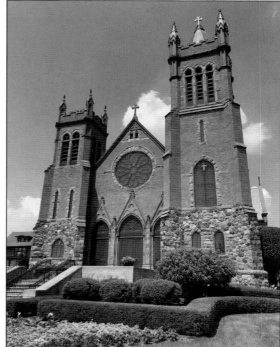

As it has for well over a century, St. Paul on the Lake still stands tall and inspires all who pass by, whether by boat, by car, or on foot. (Authors' photograph.)

Like many denominations, due to the fast-growing suburbs of Grosse Pointe, the English District Mission Board of the Lutheran Church–Missouri Synod determined that a mission should be instituted there. In February 1940, a mission was established, with the first services held in the library at Grosse Pointe High. Later that year, the new church began renting the Punch and Judy Theatre for services, as seen here. (Courtesy of St. James Lutheran Church.)

In 1940, at the first organizational meeting, the name St. James Lutheran Church of Grosse Pointe was chosen for the new church. The name was in honor of St. James the Greater, one of the 12 apostles. The initial plan to make the Punch and Judy residency a short one was interrupted by World War II, which ended nonessential building projects. (Courtesy of St. James Lutheran Church.)

It was not until June 1947 that ground was broken for the new church. It would be located on MacMillan Road in Grosse Pointe Farms and built in the American Colonial style. The tower and steeple were loosely patterned after those of Old North Church in Boston. The church dedication was held on December 5, 1948, as seen in this image. (Courtesy of St. James Lutheran Church.)

The dedication ceremony started at the Punch and Judy at 10:30 a.m. on December 5, 1947. After a short service, the congregation, led by the architect and builder, made the short walk to the newly finished church. Once there, a prayer and ribbon-cutting ceremony preceded the congregation entering the sanctuary. Pictured is St. James today, with members enjoying a jazz concert on a fine spring day. (Courtesy of Richard Allison.)

The growth of the Grosse Pointes also convinced local Methodists that they needed a church of their own. In June 1945, a down payment was made on a plot of land on Moross Road in Grosse Pointe Farms. In the meantime, the new congregation would meet at the Kerby School, which was then in a building next to the town offices. (Courtesy of Grosse Pointe United Methodist Church.)

October 14, 1945, was designated Charter Sunday. At that service, 43 people who had indicated their willingness to join the church were taken in as charter members. The group was photographed on November 4, 1945. (Courtesy of Grosse Pointe United Methodist Church.)

Grosse Pointe Methodist Church

FIRST SERVICE

ANNOUNCING

the first services here by Methodists in the history of Grosse Pointe

Sunday, September 9

in

KERBY SCHOOL

Kerby Rd., Between Grosse Pointe Boulevard and Kercheval Ave.

With
Dr. J. Adolph Holmhuber

Executive Secretary, Methodist Union of Greater Detroit

Preaching

Rev. Hugh C. White
Minister

SERVICES

SUNDAY SCHOOL for Young People - 12 years and over and adults: 10:00 A. M.

MORNING WORSHIP 11:00 A. M.

SUNDAY SCHOOL for Children under 12 and Nursery Dept. 11:00 A. M.

Church has acquired site on Moross at Ridge Road

On New Year's Day 1950, a Sunday, the cornerstone was formally laid. Bad weather had delayed the ceremony, which explains the year "1949" carved into the stone. (Courtesy of Grosse Pointe United Methodist Church.)

The first service in the brand new Grosse Pointe Methodist Church, as it was called then, was on September 17, 1950. The formal opening-day service was held on December 3, 1950, with Bishop Marshall E. Reed and district superintendent Dr. William R. Harrison officiating. This aerial photograph is from the late 1950s, and the street at upper left is Ridgemont. (Courtesy of Grosse Pointe United Methodist Church.)

The Memorial Garden, shown here during a service, was created in 1972 by volunteers from the congregation led by Rev. David Kidd. It is in the English tradition of gardens with stone walls surrounding a lower level with planted beds. The ashes of the departed are placed directly into the soil and become part of the life of the gardens. A record of those inhumed is kept in memory books in the church. (Courtesy of Grosse Pointe United Methodist Church.)

The senior high group from Grosse Pointe Methodist went on a Wesleyan heritage tour to England in the summer of 1975. They toured for four weeks, leading worship services and putting on concerts. From left to right are (first row) Ruth Davey, Blaire Houchens, Jacki van Becelaere, Liz Jewell, Laurel Shover, and Monica Bilquist; (second row) Bob Roddewig, Faye Songe, Mary Slone, Jody Nelson, Charlie van Becelaere, and David Moore; (third row) Tim Thomas, Rick Bantien, Stuart Sweet, Greg Momeyer, Doug Waugaman, and Aaron Chatterson. (Courtesy of Grosse Pointe United Methodist Church.)

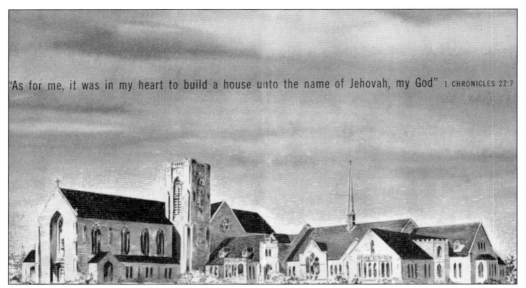

The front page of the June 27, 1957, issue of the *Grosse Pointe News* features an image nearly identical to this one. The story went on to speak of the church's hope to begin a campaign by 1959 to raise funds for an extension in order to complete the vision for the church. (Courtesy of Grosse Pointe United Methodist Church.)

Forty-three years later, the vision came to pass in 2000, when this stunning updated version opened for worship. (Courtesy of Grosse Pointe United Methodist Church.)

Grosse Pointe Congregational Church was formed on May 14, 1940, at the home of Mr. and Mrs. Huette, with 37 people in attendance. Initially, services were held at Maire Elementary School but were moved to Richard Elementary in the fall of 1942. (Courtesy of Grosse Pointe Congregational Church.)

THE GROSSE POINTE CONGREGATIONAL CHURCH

Meeting at Richard School
McKinley near Kercheval

CHARLES W. SCHEID
Pastor

Friday, 7:30 p. m.
Church Family Christmas Party at Richard School

Sunday, II a. m.
Christmas Worship Service

Sunday school attendees pose on the Richard Elementary School steps in the fall of 1949. Nice hats seem to be the order of the day. (Courtesy of Grosse Pointe Congregational Church.)

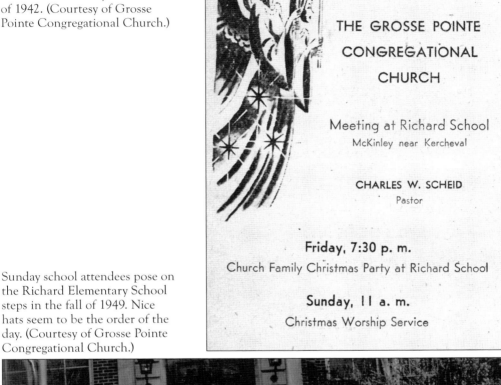

Ground breaking for the sanctuary on Chalfonte Avenue occurred on November 14, 1948, with the cornerstone being laid on March 6, 1949. By the time this picture was taken later in 1949, the building's construction was proceeding well. (Courtesy of Grosse Pointe Congregational Church.)

The dedication of Grosse Pointe Congregational Church was held on January 15, 1950. Over 1,500 people attended the event. This beautiful image is from that year. (Courtesy of Grosse Pointe Congregational Church.)

Christmas is an especially joyous time at Grosse Pointe Congregational Church. Seen here in the late 1950s are the traditional altar decorations for the holiday season. (Courtesy of Grosse Pointe Congregational Church.)

The young people of the church gather for the ground breaking of the educational and youth building. Their building was dedicated on March 27, 1955. (Courtesy of Grosse Pointe Congregational Church.)

The young man on stage seems amazed by Milky the Clown's magic skills at the church father/children banquet in 1960. Bonus points for remembering Milky's magic words! (For those who don't remember, they are "Twin Pines," the name of a now closed Detroit dairy that sponsored a TV show featuring Milky.) (Courtesy of Grosse Pointe Congregational Church.)

The Congregational youth group held a car wash in 1958. In the background is the newly opened Brownell Junior High School. (Courtesy of Grosse Pointe Congregational Church.)

In 1871, several members of Trinity Evangelical Lutheran Church in Detroit met to organize a new parish to be known as St. Paul Evangelical Lutheran Church. At this meeting, John Haaker, Ernst Glogner, and Christian Schoening were elected trustees. Shortly afterward, property at the corner of Joseph Campau and Jay Street in Detroit was purchased, and a contract for a church was made at a total cost of $10,140, around $253,000 today. (Courtesy of St. Paul Evangelical Lutheran Church.)

Under pastor Charles W. Sandrock, who served for 28 years beginning in 1948, many notable events occurred in the life of the church. In 1949, land was purchased at the corner of Lothrop Road and Chalfonte Avenue in Grosse Pointe Farms, and a new church building (seen here) was erected and dedicated in 1950. An education building was added in 1961, and the congregation grew from 600 to around 1,200 during Sandrock's stewardship. (Authors' photograph.)

By the 1950s, the number of adherents of Christian Science in Grosse Pointe warranted the need for their own church. In 1955, they were actively searching for property. In 1960, a plot on Chalfonte Avenue belonging to the nearby tennis club was made available. The result is the First Church of Christ, Scientist house of worship seen here. (Courtesy of Grosse Pointe Church of Christ, Scientist.)

COPYRIGHT, 1906, BY M. FRENCH.

BIRTHPLACE OF MARY BAKER G. EDDY.

In 1866, Mary Baker Eddy (1821–1910) experienced a dramatic recovery from a life-threatening accident after reading an account of one of Jesus's healings. From that moment, she wanted to know how she had been healed. In 1875, she wrote *Science and Health with Key to the Scriptures*, which outlined the theology of Christian Science. This book has become Christian Science's central text, along with the Bible. (Courtesy of LOC.)

CHRIST CHURCH, GROSSE POINTE, MICHIGAN

Up until the 1920s, Episcopalians in Grosse Pointe worshiped at Christ Church on East Jefferson Avenue in Detroit. In 1928, work began on Christ Church Chapel on Grosse Pointe Boulevard in the Farms. The chapel was set far back from Grosse Pointe Boulevard to allow space for future expansion. The cornerstone was laid on September 30, and Christ Church Chapel, seen in this 1940s postcard, was dedicated on December 14, 1930. (Authors' collection.)

The expansion into a cathedral did not happen, but Christ Church Chapel is a beloved landmark today. In 1957, Lanny Ross, a popular singer on radio, television, and films, made a visit to Christ Church. He had been invited by choir director Richard Roeckelein to make a special appearance with the Christ Church Choir. Ross, front and center, had been an Episcopal choir boy in New York City. This photograph was taken in the choir rehearsal room. (Courtesy of the Garland family.)

In the summer of 1928, the decision was made to establish libraries in Grosse Pointe. In December of that year, branches were opened in Grosse Pointe Woods (then called Lochmoor) and Grosse Pointe Park. The first Grosse Pointe Farms Library opened in January 1929 and was housed in the town municipal center, pictured in the second chapter. Shown here is the interior of the original Farms library around 1939. (Courtesy of GPPL.)

In June 1929, the Grosse Pointe Public Library was formally established by the school district. Today's GPPL Central Branch on Kercheval is an outstanding example of Mid-century Modern architecture. It was designed by renowned modernist architect Marcel Breuer. A Bauhaus disciple, Breuer designed several important buildings around the world including in London, New York, Paris, and Argentina. Here is Breuer's rendering of the proposed library. (Courtesy of GPPL.)

In 1951, ground was broken for the new Grosse Pointe Central Library. Taking his turn with the shovel is Carter Sales, grandson of the late Murray Sales, who had earmarked the proceeds from the sale of his home to be spent on a new library. Behind Sales are, from left to right, past school board president Franklin Dougherty, school superintendent James Bushong, and Marion Kelly, president of the Friends of the Library. In the background between Bushong and Kelly is librarian Melitta Roemer. (Courtesy of GPPL.)

The new library opened in January 1953 and has been a landmark at the corner of Fisher Road and Kercheval Avenue ever since. Carter Sales provided the opening remarks at the library's opening. Suspended from the ceiling of the main reading room is an Alexander Calder moving sculpture donated by philanthropist W. Hawkins Ferry. (Courtesy of GPPL.)

It may be a bit dicey to show the library's colorful Calder mobile in black and white, but it is still worth a look. The sculpture has been hanging in the main reading room since the library's opening. Of course, the best option is to view it in person. (Courtesy of GPPL.)

By the mid-1950s, the library had established a bookmobile. Here, the children's librarian loads up before getting behind the wheel of this 1955 Ford Ranch Wagon en route to a St. Clare school visit. (Courtesy of GPPL.)

The new building's director's office, seen here around 1954, is clean and bright with a great view. Another perk for the director is the door leading to a small balcony. (Courtesy of GPPL.)

This is a scene the director might have seen from her balcony as Grosse Pointe High students wind their way home on a spring day, sometime around 1956. (Courtesy of GPPL.)

Joining the Central Branch library as another outstanding example of modernist architecture in Grosse Pointe Farms is the Charles and Ingrid Koebel house on Cloverly Road. The home was commissioned by the Koebels after meeting architect Eliel Saarinan on a trip to Europe in the late 1930s. Finnish-born Eliel and his son Eero submitted the design for the house in 1939, and in 1940, the Koebels moved in. (Courtesy of LOC.)

The love of design ran through the Saarinan family. The Koebels' elegant dining room with pleasing curves was designed by Pipsan Saarinan Swanson, Eliel's daughter. Eliel was on staff and designed the campus for Cranbrook School in Bloomfield Hills as well as designing "the Shed" at Tanglewood, the Boston Symphony Orchestra's summer home. Eero designed GM's Tech Center in Warren, the TWA terminal at JFK Airport in New York, and the Gateway Arch in St. Louis, Missouri. (Courtesy of LOC.)

Mrs. Donald B. Leahy, Mrs. O.D. Anderson, and Mrs. William G. Power pay a visit to the Grosse Pointe War Memorial in 1950. In 1910, Russell A. Alger and his wife, Marion, moved into this home, which they called the Moorings. After Russell's death in 1930, the family looked for ways to make the house an asset to the community. In 1949, it was donated to the newly formed Grosse Pointe War Memorial Association. (Authors' collection.)

The War Memorial was dedicated to the 3,500 Grosse Pointers who served, the 126 who died in World War II, and also to all those who have served. This mission is exemplified by this 2009 ceremony in which command master chief Adrian Williams presents the national ensign to his mother, seated at the left, during his retirement ceremony at the War Memorial. Williams, a Detroit native, had ended his 32-year active-duty career at the Michigan Navy Recruiting District. (Courtesy of Ens. Kristine Volk.)

The War Memorial currently serves an estimated 250,000 people per year. More than 3,000 functions are held annually, including programs to honor veterans, engagement experiences for adults and children, and community events. Alger grandson Fred M. Alger, pictured here, made a generous donation to create the Fred M. Alger Center for Arts, Culture and Humanities at the War Memorial. (Courtesy of the Grosse Pointe War Memorial.)

The Fred M. Alger Center, pictured here, opened in November 2022. The 25,000-square-foot facility with sweeping views of Lake St. Clair features a full restoration of the 1961 crystal ballroom and a new colonnade entrance. The center also affords new spaces for reflection and community gathering. (Authors' photograph.)

Grosse Pointe did not escape the ravages of the 1918 Spanish flu epidemic. Many cases went untreated due to a lack of local medical care. In 1919, three Grosse Pointe women— Anna Thompson Dodge, Maud Ledyard Van Ketteler, and Romayne Latta Warren—worked together to create a facility in Grosse Pointe that could treat those who needed care. The original facility, pictured here, was a home in Grosse Pointe Farms on Oak Street, now Muir Road, with room for 13 patients. That building no longer exists. (Courtesy of HFH, LAM Archives.)

The concept of cottage hospitals largely came out of England in the early 19th century. The idea was to use small rural buildings that could provide medical care without patients having to travel a long distance. Cottage Hospital co-founder Baroness Maud Ledyard Van Ketteler, pictured here, was the daughter of Grosse Pointe railroad magnate Henry Brockholst Ledyard Jr. and Mary L'Hommedieu. In 1897, she married Baron Clemens Van Ketteler of Germany, gaining her title. (Courtesy of HFH, LAM Archives.)

As the hospital prepared for its 1919 opening, a visiting nurse program was put in place. Here, nurse Johanssen heads out on a visit in 1918. (Courtesy of HFH, LAM Archives.)

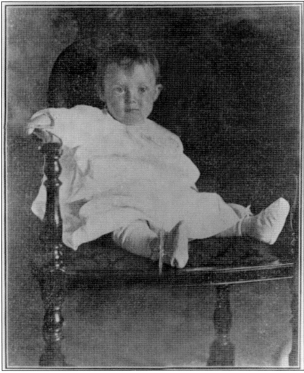

The first baby born at Cottage was Maud Van Ketteler Bremer, on June 19, 1919. She was named in honor of the co-founder of the hospital. (Courtesy of HFH, LAM Archives.)

ENTRANCE ELEVATION

Plans on Back

END AND REAR ELEVATIONS
COTTAGE HOSPITAL, GROSSE POINTE, MICH.
STEVENS & LEE, ARCHITECTS

849

With the flu epidemic at an end and Cottage beginning to gain patients from accidents, surgeries, and childbirths, the need for a larger facility became apparent. Several prominent Grosse Pointe families, such as the Newberrys and Joys, chipped in for the new hospital. The new Cottage Hospital opened in November 1928. According to *Architectural Forum*, the building cost, without equipment, was $266,000, about $4.5 million today. With a capacity of 50 patients, that worked out to be $5,300 per bed ($90,000 today). (Courtesy of the University of Michigan.)

Another of the three women who founded Cottage was Anna Thompson Dodge, wife of Horace Dodge. She is pictured on the left with her daughter Delphine in 1920. Cottage Hospital was totally women-run until 1962, when men were appointed to the board of trustees. (Courtesy of the Church of Latter Day Saints.)

This is Cottage Hospital around 1940. By then, the white paint seen in previous images was gone. (Courtesy of HFH, LAM Archives.)

This image provides a good look at the hospital lab in the 1940s. Cottage Hospital today is officially known as the Henry Ford Medical Center–Grosse Pointe and occupies several blocks on and around Kercheval Road in Grosse Pointe Farms on the north end of the Hill district. (Courtesy of HFH, LAM Archives.)

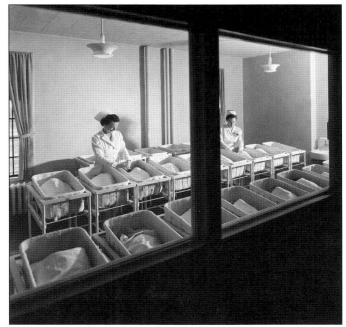

As a result of eating tainted olives at a dinner party given by Grosse Pointe residents Murray and Jessie Sales in 1920, several guests fell seriously ill. Five people died of food poisoning, including two Sales children. As a memorial to them, in 1926, the Saleses donated $100,000 ($1.7 million today) for a children's wing at the new Cottage Hospital. Shown here in happier times, two nurses tenderly watch over their charges in the hospital maternity ward in the 1960s. (Courtesy of HFH, LAM Archives.)

DISCOVER THOUSANDS OF LOCAL HISTORY BOOKS FEATURING MILLIONS OF VINTAGE IMAGES

Arcadia Publishing, the leading local history publisher in the United States, is committed to making history accessible and meaningful through publishing books that celebrate and preserve the heritage of America's people and places.

Find more books like this at
www.arcadiapublishing.com

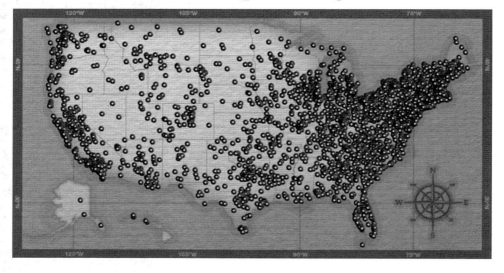

Search for your hometown history, your old stomping grounds, and even your favorite sports team.